Encyclopedia of Earthquake Research and Analysis: Overview
Volume III

Edited by **Daniel Galea**

New York

Published by Callisto Reference,
106 Park Avenue, Suite 200,
New York, NY 10016, USA
www.callistoreference.com

**Encyclopedia of Earthquake Research and
Analysis: Overview
Volume III**
Edited by Daniel Galea

International Standard Book Number: 978-1-63239-236-7 (Hardback)

Printed in the United States of America.

Contents

Preface

To minimize human and material losses due to inevitable occurrence of earthquakes, their study and understanding is of utmost importance. Some of the many topics included in this book are statistical seismology studies, the latest techniques and advances on earthquake precursors and forecasting, and also, new ways for immediate detection, data acquisition and its interpretation. The book covers a wide range of topics, from theoretical advances to practical applications.

The information contained in this book is the result of intensive hard work done by researchers in this field. All due efforts have been made to make this book serve as a complete guiding source for students and researchers. The topics in this book have been comprehensively explained to help readers understand the growing trends in the field.

I would like to thank the entire group of writers who made sincere efforts in this book and my family who supported me in my efforts of working on this book. I take this opportunity to thank all those who have been a guiding force throughout my life.

Editor

Part 1

Earthquake Observatories

The Effect of Marmara (Izmit) Earthquake on the Chemical Oceanography and Mangan Enrichment in the Lower Layer Water of Izmit Bay, Turkey

Nuray Balkis
Istanbul University, Marine Science and Management Institute, Istanbul
Turkey

1. Introduction

Dissolved oxygen (DO) content of the marine environment is a crucial parameter for life and water quality as well as playing an important role in biogeochemical processes, and respiration of plants and animals, and decomposition of organic matter by bacteria are the primary processes that consume dissolved oxygen content of water and pore-water in sediments. If the oxygen concentration of water falls below about 2 mg/1, living organisms become stressed and the consequent conditions lead to hypoxia. Persistent hypoxia and increased oxygen uptake accompanies release of hydrogen sulfide. Anoxia occurs in estuaries where high loads of organic matter and/or nutrients are supplied, and in semienclosed water bodies where water mixing and tidal exchanges are strongly restricted.

In recent years, aquatic ecosystems have been contaminated by heavy metals; which are of agricultural, industrial, domestic, mining and also natural origins (Ayas and Kolankaya 1996; Han et al., 2002). They are potentially toxic to the aquatic environment; if they exceed natural limits, they will be harmful to the aquatic organisms' environments and human health (Forstner and Witmann, 1981). Organisms need some metals such as Fe, Cu, Zn, Co, Se, Ni and Mn in certain amounts; however, exceeding these amounts may cause toxic effects for these organisms. Some metals such as Hg, Cr, Pb and Cd are toxic to organisms and marine habitat. These metals are dissolved in sea water or suspended in solid materials and absorbed through the gills or skin of marine organisms; they also accumulate in the bodies of organisms through the food chain (Forstner and Witmann, 1981). Mussels, in particular, have been used as biological indicator organisms to monitor marine pollution by toxic heavy metals and potentially toxic chemicals due to their properties of inhabitation (Pempcowiac et al., 1999; Hu 2000).

Izmit Bay (Figure 1) is a semi-enclosed water body and situated in the NE of Marmara Sea. It has been subjected to pollution problems (Orhon et al., 1984; Tuğrul et al., 1989; Morkoç et al., 1996), including eutrophication of the water and inputs of toxic industrial and domestic effluents. Total organic matter load of industrial discharges has been reduced to 80% within the last 10 years, whereas domestic organic loads have been increased in two fold (Morkoç et al., 1996, 2001). The earthquake with a magnitude of 7.4 was occurred at 17th of August 1999, destroying the eastern Marmara Region. The epicenter of the earthquake was found to be in a small city (Gölcük) located on the southern coast of Izmit Bay. This seismic event caused the destruction of wastewater discharge systems and also dispersal of refined petroleum products onto the sea surface from the subsequent refinery fire. The surface waters of the Bay were partly covered by the thick petroleum layers and partly by a film (Güven et al., 2000, Ünlü et al., 2000). Petroleum layer covering the surface water reduced the transfer of oxygen from air/sea interface and also caused the subsequent death of living organisms. Increasing effluent discharges into the Bay produced an exceptional plankton bloom. Coupling of such factors leading to oxygen deficiency at the sea floor caused the formation of anoxic conditions. Okay et al., (2001) investigated ecological changes in Izmit Bay, however their data is limited with the September 1999.

This paper presents the results of one-year monitoring program performed in Izmit Bay after the Earthquake, with the purposes of describing the abrupt changes in chemical oceanography and understanding the mechanism of H_2S generation in the Bay which has not been occurred before. Furthermore, the factors controlling metal distributions in water column and surface sediments of the Bay were discussed in this study.

1.1 Study area

Izmit Bay is an elongated semi-enclosed water body with a length of 50 km, width varying between 2 to 10 km (Figure 1) and has an area of 310 km^2. The bathymetry of the Bay constitutes three sub-basin separated by shallow sills from each other. The eastern basin is relatively shallow (at about 30 m) whereas the central basin has two small depressions with depths of 160 and 200 m. The western basin deepens in westward from 150 m to 300 m and connects the Bay to the Marmara Sea. Izmit Bay is oceanographically an extension of Marmara Sea, having a permanent two-layered water system. The upper layer is originated from less saline Black Sea waters (18.0-22.0 psu), whereas the lower layer originated from the Mediterranean Sea waters is more saline (37.5-38.5 psu) (Ünlüata et al., 1990). The permanent stratification occurs at about 25 m in the Marmara Sea (Beşiktepe et al., 1994), however it is highly variable in Izmit Bay (Oğuz and Sur, 1986) (Figure 2). The thickness of the upper layer changes seasonally from 9 to 18 m spring and autumn, respectively (Oğuz and Sur, 1986; Algan et al., 1999). The upper layer enters into the Bay in spring and summer, corresponding to the freshwater inflow changes in the Black Sea, while the lower layer flows to the Marmara Sea from the Bay. However, the upper layer flows towards the Marmara Sea in autumn and winter (Oğuz and Sur, 1986). Vertical mixing of the two layers is restricted and occurs at shallow depths. An intermediate layer develops throughout the year in the water column of the Bay with varying thickness (DAMOC, 1971; Baştürk et al., 1985; Tuğrul et al., 1989; Oğuz and Sur, 1986; Altıok et al., 1996). The upper layer of Izmit Bay, in general, is saturated with DO (Tuğrul and Morkoç, 1990). DO concentrations in the lower layer of Izmit Bay has been found to be 2.5-3.0 mgl^{-1} in winter and spring periods and 0.7-1.5 mgl^{-1} in summer, in previous studies (Morkoç, et al., 1996). The minimum DO concentrations have been measured locally in the central basin (0.1-0.2 mgl^{-1}) and in the eastern basin (0.5 mgl^{-1}) during spring-summer period (Tuğrul and Morkoç, 1990). Izmit Bay and its surroundings is one of the most industrialized and populated area of Turkey, receiving more than 300 industrial and domestic effluents (Morkoç et al., 1996). Industrial effluents discharges a total of 163,000 m^3/day wastewater, 24 tons/day BOD and 19,5 tons/day TSS to Izmit Bay (Morkoç et al., 2001). The eastern basin receives the highest inputs compare to other basins of the Bay. Based on the previous studies, no DHS has been measured in Izmit Bay (Morkoç et al., 1988; Tuğrul et al., 1989; Morkoç et al., 1996).

The Effect of Marmara (Izmit) Earthquake on the Chemical Oceanography and Mangan Enrichment in the Lower
Layer Water of Izmit Bay, Turkey

5

Industrial loads have been reduced by treatment and waste minimization within the last 10 years, but domestic wastes has doubled, due to the increasing population in the Bay. Therefore, the total (domestic + industrial) discharge load into the Bay during the last 10 years has not changed significantly (Morkoç, et al., 2001). The dissolved oxygen content of Izmit Bay decreased dramatically from 1984 to 1999 and reached to a minimum value at 20 m throughout the Bay (Okay et al., 2001). Ergin et al., (1991) suggested that the surface sediments in İzmit Bay are uncontaminated by anthropogenic pollution. However Yaşar et al., (2001) investigated that the heavy metal concentrations are highest in the eastern and central basins. The western basin was also found generally unpolluted with respect to heavy metals by Yaşar et al., 2001.

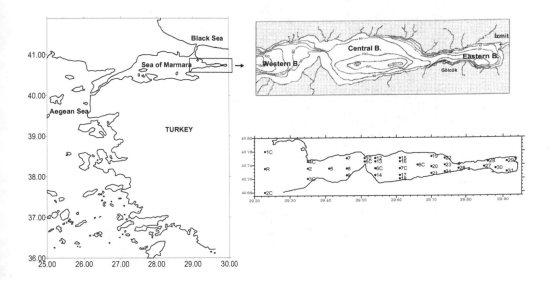

Fig. 1. The location (left) and bathymetry (above) of the study area. The location of sampling station in İzmit Bay (below).

2. Material and methods

2.1 Sampling sites

The water samples were collected from 32 stations in İzmit Bay, including one station located off the western basin (R), on board the R/V Arar (Figure 1). Station (R) represents the characteristics of the Marmara Sea and hence, provides a comparison between the Bay and the Marmara Sea. The sampling stations in İzmit Bay represent the various depths of three basins, with a minimum of 17 m and maximum of 200 m. Sampling was carried out with a Rosette sampler assembled to the Sea Bird CTD System at about 10 m depth intervals through the upper and the lower layers. Sampling period includes August 1999, immediately after the Earthquake and performed monthly in 1999 and in February, May and August during 2000.

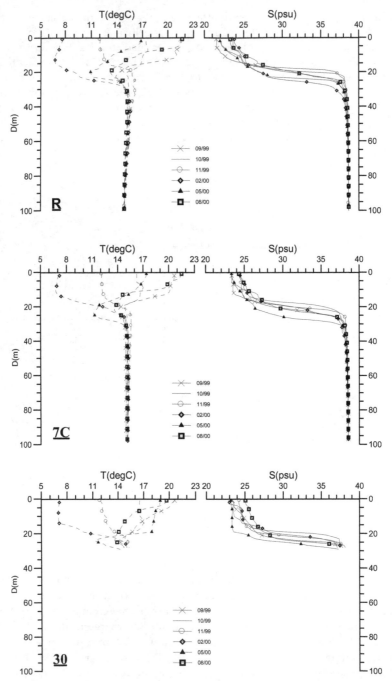

Fig. 2. Depth profile of salinity and temperature distribution along the water column of some selected station from İzmit Bay (from Güven et al., 2000).

The Effect of Marmara (Izmit) Earthquake on the Chemical Oceanography and Mangan Enrichment in the Lower
Layer Water of Izmit Bay, Turkey

7

2.2 Analytical methods

Samples for DO determinations were drawn first from the Niskin bottles of Rosette to prevent any biological activity and gas exchanges with the atmosphere. DO determination was carried out by Winkler method (Greenberg et al., 1985) on board from all the stations (Figure 1). The precision of method was estimated at ± 1.9 % .

Dissolved hydrogen sulfide (DHS) was measured only at stations where DO concentrations are lower than the detection limit of the method (0.03 mgl⁻¹) (Figure 1), in all the sampling periods, except August and September 1999. DHS contents were measured by an iodometric titration method (Strickland and Parsons, 1972).

pH values measured along the water column at all stations with a WTW 526 pH-meter equipped with a temperature-compensation adjustment on board.

The water samples were filtered through 0.45 µm filters using metal clean tecniques (Bruland et al., 1979). The samples were stored in polyethylene bottles (LDPE) that were acid cleaned using methods described Patterson and Settle (1976). After collection, the samples were acidified to a pH between 1.5 and 2.0 using HNO_3. Dissolved heavy metal concentrations (Fe, Mn, Pb, Cu and Cd) were measured by atomic absorption spectrophotometer (AAS) following preconcentration with ammonium 1-pyrrolidinedithiocarbamate (APDC) in an organic extraction (Bruland et al., 1985). The blanks for the metals analyzed were: Fe, 010 ± 0.05 mg/l; Mn, 0.10 ± 0.02 mg/l; Cu, 0.15 ± 0.08; Cd, 0.05 ± 0.03 mg/l; Pb, 0.20 ± 0.30 mg/l.

The surface sediments total carbonate contents were determined by a gasometric-volumetric method (Loring & Rantala, 1992). Total organic carbon (C_{org}) was analyzed by the Walkey-Blake method, which involves titration after a wet combustion of the sample (Gaudette, 1974; Loring & Rantala, 1992). Al, Fe, Mn, Cu, Zn, Co and Cr contents were determined by atomic absorption spectrophotometer (AAS) after a "total" digestion, involving HNO_3 + $HClO_4$ + HF acid mixture.

The sequential selective extraction analyses were carried out using 1M Na-acetate (pH=5 adjusted by acetic acid) for the dissolution of carbonate phase, 0.04M hydroxylamine hydrochloride (HAHC) in 25% acetic acid for dissolving Fe-Mn-oxyhydroxides, 0.02M nitric acid + 30% hydrogen peroxide (pH=2) for extracting organic matter, and HNO_3 + $HClO_4$ + HF mixture for the total extraction of the residual (lithogenous) fraction (Tessier et al., 1979).

Reference Material	Element	Measured value (this study) ppm	Certified value or range ppm
SL-1	Fe	62	65-7-69.1
IAEA405	Al	63500	72700-83100
SL-1	Cr	98	95-113
SL-7	Mn	634	604-650
CRM-142	Cu	25	27.5
CRM-142	Zn	92	92.4
CRM-142	Co	13	7.9

Table 1. Accuracy of ASS analyses used in this study, as determined by Analysis of AQCS (SL-1 and SL-7), IAEA405 and BCR (CRM 142) reference materials.
SL-1 and SL-7 are lake sediment and CRM 142 is a light sandy soil.

The precision of the "total" metal analyses and the different selective extraction steps is better than 10% and 15%, respectively, at 95% significance level. The accuracy of the analyses were checked by analyzing the AQCS (lake sediment SL-1 and SL-7), IAEA405 and BCR (light sandy soil CRM 142) reference materials (Table 1). The metal values were normalized to eliminate the grain-size effects using metal/Al ratios (e.g., Loring & Rantala, 1992).

3. Results

3.1 Dissolved Oxygen (DO)

The vertical changes of DO in water column are shown by selecting some of the stations from each of the three (western-central-eastern) basins where the variations are significant (Figure 3). In station (R), DO concentrations are between 7.4-10.7 mgl^{-1} for the upper layer, and 1.1-1.5 mgl^{-1} in the lower layer during the sampling period between August 1999 to August 2000 (Figure 3), which are similar to vertical DO concentrations of the Marmara Sea (Ünlüata and Özsoy, 1986; Ünlüata et al., 1990; Doğan et al., 2000). Depth profile of DO concentration in İzmit Bay displays a sharp decrease at about 20 m below the surface in the western and the central basins, following the water stratification during late summer and autumn (Güven et al., 2000) (Figure 3). In February the gradual decline of DO occurs at about 30 m water depth.

DO concentration of the water column in the eastern basin indicates a more gentle profile, probably due to the vertical mixing of the two layers (Oğuz and Sur, 1986; Altıok et al., 1996) in shallow depths. DO concentration of the upper layer in İzmit Bay varies in a range between 4.5 to 12.1 mgl^{-1} during August 1999-2000 (Figure 4a and b). Saturated DO occurs in the upper layer locally at the eastern basin in October 1999 (Figure 4a, 5). The saturation concentration of DO (SDO) values are determined by the solubility oxygen in sea water as a function of the concurrently measured values of temperature and salinity (Figure 2, 5 and 6). The highest DO concentrations were measured in February 2000 (Figure 4b), whereas the lowest DO concentrations were measured in September 1999 after the earthquake. The distribution of DO concentration in the upper layer is almost homogeneous in August and September 1999. From October 1999 to February 2000, DO concentration of the upper layer significantly increases. The increase of DO concentration occurred subsequently following the maximum phytoplankton bloom in October (Güven et al., 2000). DO concentrations range between 7.9-14.5 mgl^{-1} in October 1999, displaying the highest value in the eastern basin (Figure 4a and 5). The distribution of DO concentration in the upper layer becomes almost homogeneous with a range of 9.0-10.4 mgl^{-1} in December 1999 and slightly increases (10.19-12.12 mgl^{-1}) in February 2000. DO concentrations in the upper layer decrease (6.69-10.57 mgl^{-1}) in spring (May-2000, Figure 4b) and are lower in the eastern basin compare to the central and the western basins of İzmit Bay. In August 2000, DO concentrations lie between 5.0 and 9.4 mgl^{-1}, being relatively high in the central basin.

DO content of the lower layer is significantly lower than that of the upper layer, throughout the sampling period (Figure 3). In August 1999, after the earthquake DO concentrations of the lower layer ranges between 0.0-1.46 mgl^{-1}. DO concentration below the detection limits (<0.03 mgl^{-1}) was measured in areas where the water depth is deeper than 100 m and also in the eastern basin (Figure 7a and b). The lowest DO concentration ranges were measured in September 1999. The deficiency of DO is mostly accompanied with the presence of DHS and

The Effect of Marmara (Izmit) Earthquake on the Chemical Oceanography and Mangan Enrichment in the Lower
Layer Water of Izmit Bay, Turkey

9

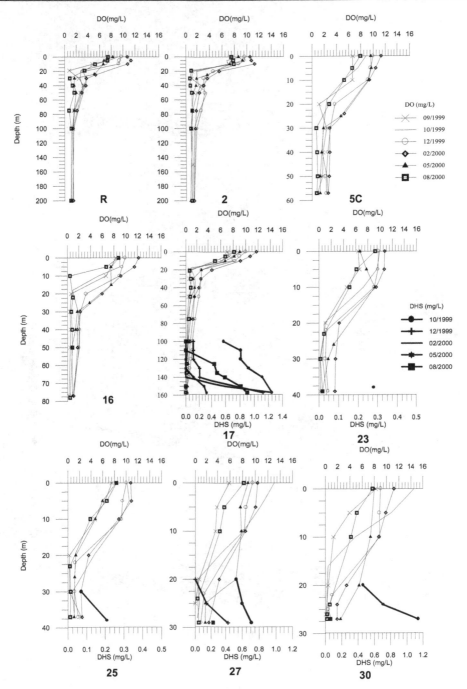

Fig. 3. Vertical distribution of DO and DHS along the water column in various stations of
İzmit Bay.

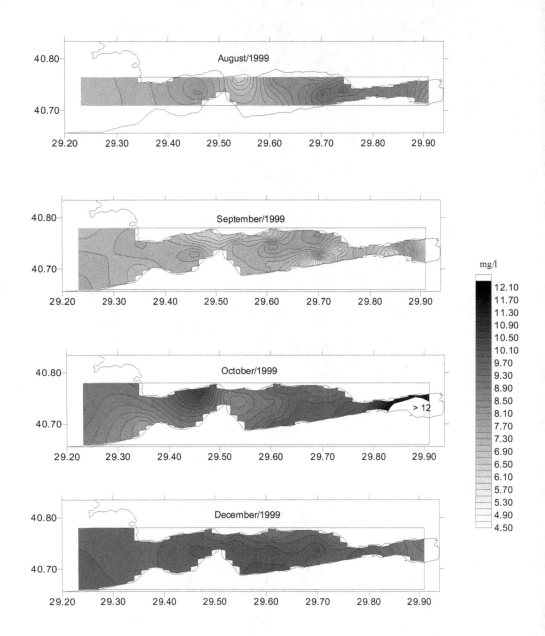

Fig. 4a. Spatial distribution of DO (mgl⁻¹) in the upper layer of İzmit Bay during August to December 1999.

The Effect of Marmara (Izmit) Earthquake on the Chemical Oceanography and Mangan Enrichment in the Lower
Layer Water of Izmit Bay, Turkey

11

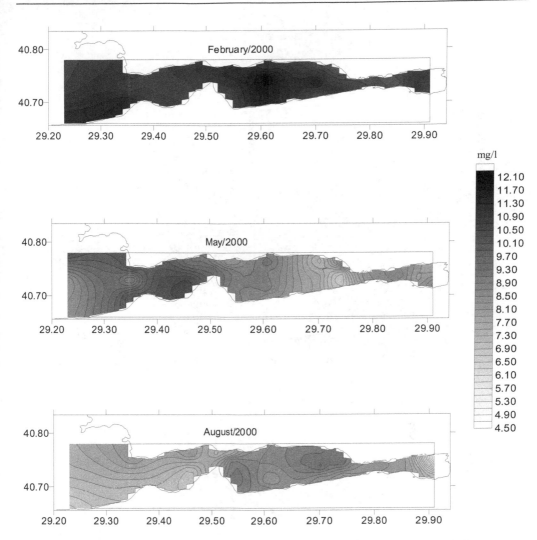

Fig. 4b. Spatial distribution of DO (mgl-1) in the upper layer of İzmit Bay during February to August 2000.

is discussed below. The local anoxic conditions continue in September and October 1999 (Figure 7a), mainly in the central basin with DO concentration range of 0.0-3.4 mgl-1 and 0.0-7.9 mgl-1, respectively. In December 1999, DO concentration of the lower layer generally increases (0.0-8.3 mgl-1), but in only one station (17) no DO was lower than the detection limits (Figure 7a). The distribution pattern of DO concentration in the lower layer displays similar behavior in February and May 2000 (Figure 7b) with ranging between 0.0-8.2 mgl-1 and 0.0-8.8 mgl-1, respectively. Relatively higher DO concentrations are close to the northern coast of İzmit Bay (Figure 7a and 7b). DO concentration of the lower layer decrease in August 2000, ranging between 0.0-3.6 mgl-1.

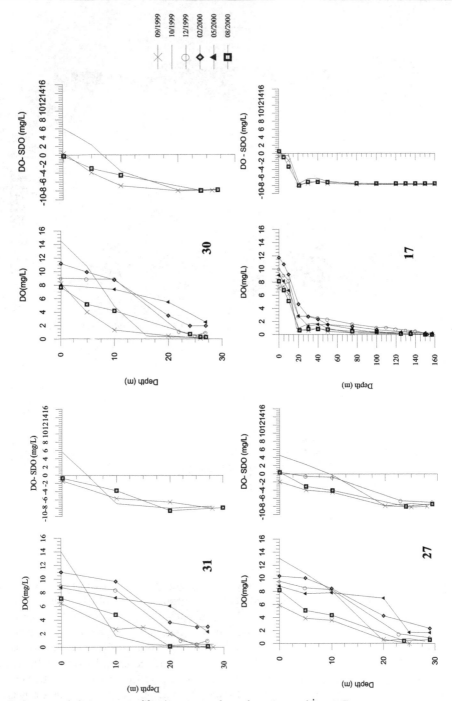

Fig. 5. Oxygen deficiency profiles in some selected stations of İzmit Bay.

The Effect of Marmara (Izmit) Earthquake on the Chemical Oceanography and Mangan Enrichment in the Lower
Layer Water of Izmit Bay, Turkey

13

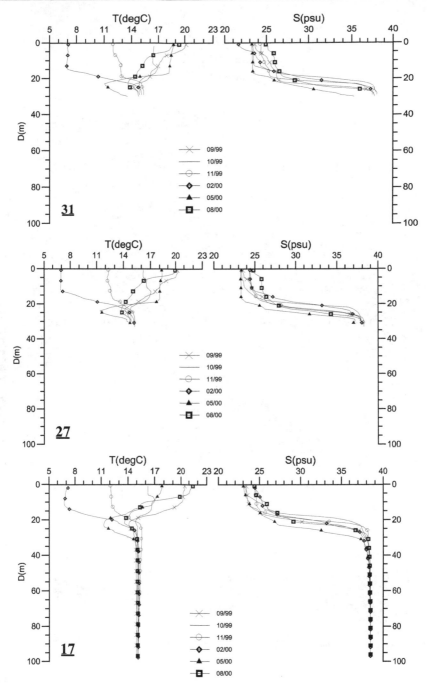

Fig. 6. Depth profile of salinity and temperature distribution along the water column of
some selected stations from İzmit Bay (from Güven et al., 2000).

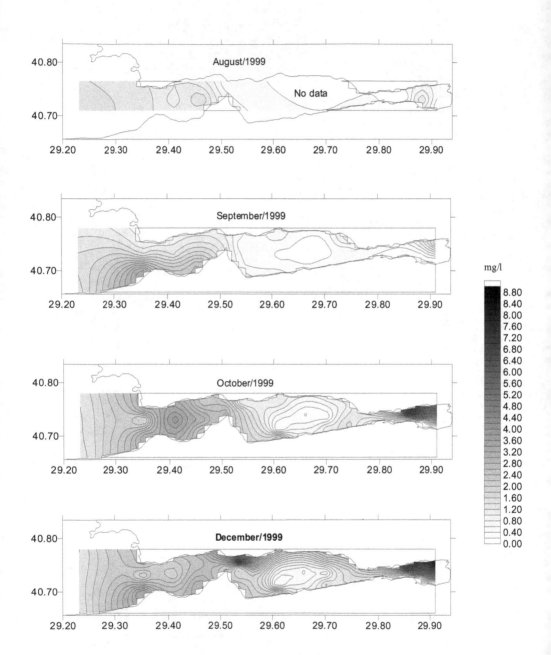

Fig. 7a. Spatial distribution of DO (mgl⁻¹) in the lower layer of İzmit Bay during August to December 1999

The Effect of Marmara (Izmit) Earthquake on the Chemical Oceanography and Mangan Enrichment in the Lower
Layer Water of Izmit Bay, Turkey

15

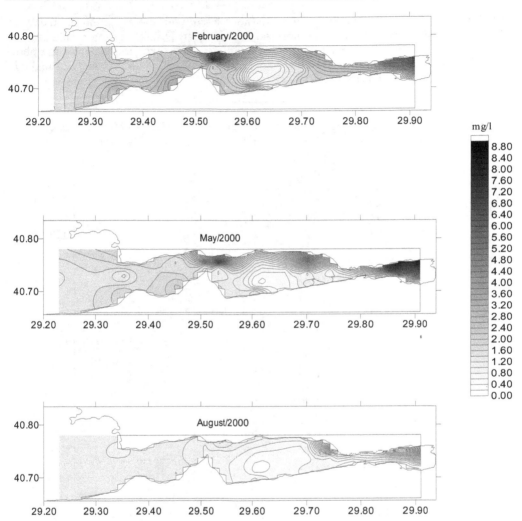

Fig. 7b. Spatial distribution of DO (mgl⁻¹) in the lower layer of İzmit Bay during February to
August 2000.

3.2 Dissolved Hydrogen Sulfide (DHS)

The presence of DHS is limited with certain stations and in the lower layer. DHS formation
develops both in the shallow area of the eastern basin (maximum 30 m) and relatively deep
areas of the central basin. DHS is detected in stations 27, 29, 30 and 31 in the eastern basin,
and 8C, 17, 19, 20, 23 and 24 in the central basin (Figure 1, Table 1). In the station 25 situated
at the connecting small strait between the eastern and the central basin, measurable DHS
exists in the lower layer. DHS concentrations vary in between 0.06-1.25 mgl⁻¹, reaching to
their maximum in the lower layer of station 17. No DHS data available for the sampling
period of August 1999 after the Earthquake. DHS concentration in October 1999 reaches to
1.14 mgl⁻¹ in the eastern basin (station 30) (Figure 3). In general, DHS appears at a water

depth of 20 m in the shallow parts of the eastern basin, with the exception of station 29 where DHS exists up to 10 m to the surface. In December, DHS formation occurs only in stations 27 and 29 and disappears in the other stations. During winter and spring sampling period, no DHS is detected in the eastern basin. Low DHS concentration (0.06 and 0.23 mgl⁻¹) is found in stations 27 and 30 in August 2000.

The central basin							
Station	Water depth (m)	Sampling depth (m)	October 1999	December 1999	February 2000	May 2000	August 2000
25	40	30	0.07	*	*	*	*
		38	0.21	*	*	*	*
27	31	20	0.53	*	*	*	*
		25	0.60	0.14	*	*	*
		29	0.71	0.40	*	*	0.23
29	18	10	0.43	*	*	*	*
		15	0.50	*	*	*	*
		17	0.85	0.24	*	*	*
30	29	20	0.45	*	*	*	*
		24	0.71	*	*	*	*
		27	1.14	*	*	*	0.06
31	30	20	0.28	*	*	*	*
		25	0.28	*	*	*	*
		27	0.36	*	*	*	*
The eastern basin							
17	160	100	0.57	0.11	*	*	*
		110	0.81	0.11	*	*	*
		120	0.81	0.11	*	*	0.43
		130	0.92	0.21	*	*	0.48
		140	1.11	0.21	*	0.11	0.59
		150	-	-	0.60	0.28	0.81
		152	-	-	0.89	-	-
		157	1.25	1.12	1.25	0.32	0.90
8C	116	100	0.64	*	*	*	*
		114	0.93	*	*	*	*
19	29	25	0.40	*	*	*	*
20	106	75	0.28	*	*	*	*
		80	0.40	*	*	*	*
		90	0.43	*	*	*	*
		102	0.71	0.32	*	*	*
23	43	38	0.28	*	*	*	*
24	57	53	0.28	*	*	*	*

(*) No detectable DHS

Table 2. DHS concentrations in the lower layer of the eastern and the central basins of İzmit Bay.

The Effect of Marmara (Izmit) Earthquake on the Chemical Oceanography and Mangan Enrichment in the Lower
Layer Water of Izmit Bay, Turkey

17

DHS is continuously detected in station 17 from the central basin in all the sampling periods, being highest in October 1999 and lowest in May 2000. The occurrence of DHS corresponds to depth interval of 100-160 m (bottom) during October and December 1999. In February 2000, DHS concentration is limited with the lower 10 m water column, increasing to 1.25 mgl-1 at the bottom. The thickness of DHS formation layer slightly increases to 20 m in May 2000, however the concentrations are reduced compare to February 2000 (Figure 3, Table 2). In August 2000, both concentration and thickness of DHS formation of the lower layer in station 17 increase. In spite of continuous presence of DHS in station 17, it occurs solely in October 1999 in the other stations of the central basin, and never reaches to 1 mgl-1 (Figure 3, Table 1). DHS is detected mainly at the very close to bottom (2 m above the sea floor) of the lower layer in most of the stations in the central basin, however the lower 25 m and 15 m of the water column in station 20 and 8C, respectively, contain DHS.

3.3 pH
The highest pH values were measured in October 1999 and May 2000 throughout the water column in İzmit Bay, and were particularly high in the eastern basin (Figure 8 and 9). The lowest pH values, on the other hand, correspond to sampling period of August 2000. pH values significantly decrease from 8.5 to 7.1 mainly in the eastern basin and to a lesser degree in the central basin, during the formation of DHS.

3.4 Metals
3.4.1 Water column
Iron concentrations range between <4 mg/l and 21 mg/l along the water column in İzmit Bay (Table 3). The highest values are measured after the Earthquake (October-1999). High dissolved Fe concentrations indicate reduction of Fe-oxides by bacteria during mineralization of organic carbon in the sediment and diffusion into bottom waters (Nealson, 1982; Lovley and Phillips, 1988; Nealson and Myers, 1990). Fe values are decrease in May and August 2000 where Fe limitation is thought to control phytoplankton productivity.

Manganese concentrations vary between <1 and 123 mg/l in water column of the Bay (Table 3). The values increased in lower layer water and near the sediment-water interface in eastern and central basins. This was attributable to the degradation of settling organic carbon (Nealson, 1982; Nealson and Saffarini, 1994; Nealson and Myers, 1990). Manganese oxides were reduced to dissolved Mn^{+2}, which diffused from the sediment into the water column occurring the anoxic conditions. The lowest Mn values are obtained in December 1999 and February 2000. In these periods, oxygen-rich waters of Marmara Sea (Mediterranean originating) flow into the Bay. Thus, Mn-oxides are occurred and flocculated in water column with reoxidation of dissolved Mn in more oxygenated waters.

Lead concentrations range between <0.8 and 1.8 mg/l in the Bay waters (Table 3). The highest values are suggested that atmospheric and anthropogenic inputs.

Copper concentrations vary between <0.4 and 7.4 mg/l along the water column of the Bay (Table 3). The high values shows that Cu was mainly affected by redox reactions involving Mn and Fe in bottom waters of the eastern and central basins. The lowest Cu concentrations are measured in occurring the extreme phytoplankton blooms periods especially in these regions.

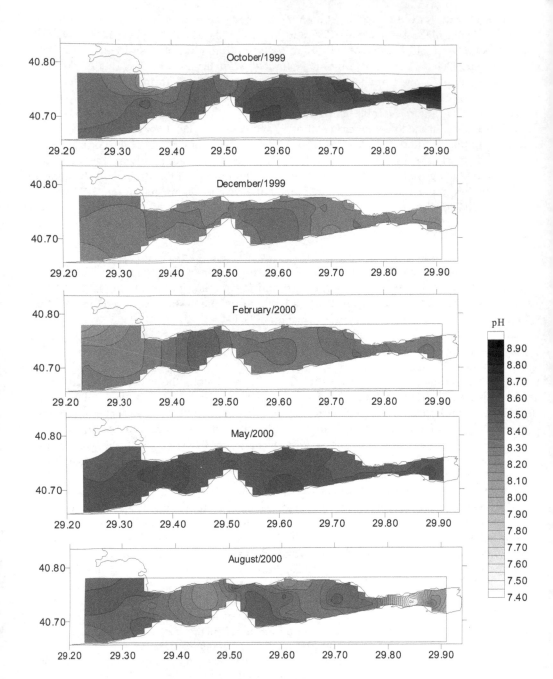

Fig. 8. Spatial distribution of pH values in the upper layer of İzmit Bay.

The Effect of Marmara (Izmit) Earthquake on the Chemical Oceanography and Mangan Enrichment in the Lower
Layer Water of Izmit Bay, Turkey

19

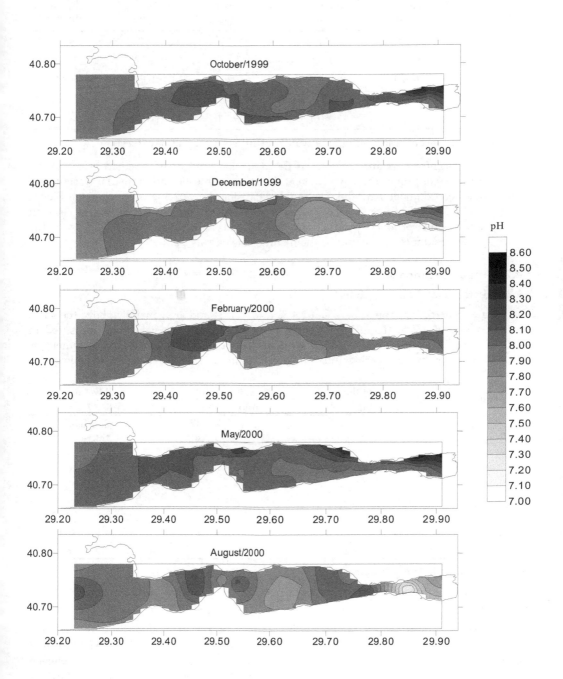

Fig. 9. Spatial distribution of pH values in the lower layer of İzmit Bay.

Element	October 1999	December 1999	February 2000	May 2000	August 2000
Fe	7-15	<4-4	<4-13	<4	<4
Mn	<1-4	1-7	2-4	4-12	<1-13
Pb	<0.8-1	<0.8-0.9	0.9-1	<0.8-2	<0.8-1
Cu	0.5-0.7	0.5-0.9	0.4-0.8	<0.4-0.6	<0.4-0.8
Cd	<0.1	<0.1	<0.1	<0.1	<0.1

Table 3. Metal concentrations along the water column of the Izmit Bay ($\mu g/l$).

Cadmium concentrations are lower than the detection limit of the method (<0.01 mg/l) along the water column of the Bay (Table 3). Since the domestic and industrial waste-water system has been damaged by the earthquake, causing the extreme phytoplankton bloom (Okay et al., 2001). This element is incorporated into organic matter by phytoplankton during periods of primary production (Sunda and Huntsman, 1995). Therefore, the relatively low residence time could be the result of biological uptake.

3.4.2 Surface sediments

3.4.2.1"Total" metal distributions

The Iron concentrations range between 2.4 % and 11.8 % and are generally above the shale average value of 4.7 % (Krauskopf, 1979) (Table 4). The highest values are measured in southern shelf and eastern basin of the İzmit Bay. The Fe distribution in the Bay sediments is controlled mainly by the riverine and anthropogenic inputs on this land-locked system.

Element	Average shale (Krauskopf, 1979)	Gulf of Izmit min - max	Gulf of Izmit mean - SD
Cu ($\mu g/g$)	50	11- 42	23 ± 8.87
Zn ($\mu g/g$)	90	84 - 306	149 ± 57
Fe (%)	4.7	4.6 - 7.1	6.1 ± 0.6
Mn ($\mu g/g$)	850	139 - 494	327 ± 89
Co ($\mu g/g$)	20	6 - 20	12 ± 3.93
Cr ($\mu g/g$)	100	34 - 77	58 ± 11
Al (%)	9.2	2.3 - 11.4	7.4 ± 2.5
CaCO3	6.0[a]	13 -42	13.4 ± 9.9
Corg (%)	0.8[a]	0.6 - 6.2	3.0 ± 1.6

[a] From Mason and Moore (1982, p.153)

Table 4. Range of metal concentrations of bulk surface sediments from different parts of the Marmara Sea

Manganese concentrations are in general, lower than the average abundance of this element in shale (<850 $\mu g/g$). The values increase in western basin of the Bay. Here, Mn^{+2} form of this redox sensitive element derived from the early diagenesis of the sediments, is believed to have been oxidized and flocculated by the oxygen-rich lower layer waters of the Marmara Sea (Mediterranean originating).

The Effect of Marmara (Izmit) Earthquake on the Chemical Oceanography and Mangan Enrichment in the Lower
Layer Water of Izmit Bay, Turkey

21

The Copper, Cobalt and Chromium concentrations are in general, below the shale average values of 50, 20 and 100 µg/g (Krauskopf, 1979) (Table 3). The highest values in eastern basins surface sediments shows that the anthropogenic inputs from the industrialized regions in here. The Cu values show high correlation with the C_{org} content (r=0.81, Table 4).

Element	Al (%)	Fe (%)	Mn (µg/g)	Cu (µg/g)	Zn (µg/g)	Co (µg/g)	Cr (µg/g)	CaCO3 (%)	Corg (%)
Al (%)	1	+0.22	-0.13	+0.17	+0.17	+0.39	+0.30	-0.57	+0.1
Fe (%)		1	-0.03	+0.37	+0.44	+0.14	+0.54	- 0.32	+0.22
Mn (µg/g)			1	-0.28	-0.32	-0.12	+0.08	+ 0.1	-0.20
Cu (µg/g)				1	+0.89	+0.55	+0.66	- 0.39	+0.81
Zn (µg/g)					1	+030	+0.61	-0.28	+0.78
Co (µg/g)						1	+0.39	-0.56	+0.46
Cr (µg/g)							1	-0.59	+0.55
Corg (%)								1	-0.1
CaCO3 (%)									1

Table 5. Corelation coefficient of matrix geochemical parameters of sediments.

Zinc concentrations range between 84 µg/g and 306 µg/g and are above the shale average value of 4.7 % (Krauskopf, 1979) (Table 4). The high values seem to have been controlled mainly by the anthropogenic inputs from the eastern region similar to the other elements. This element shows a high correlation with the C_{org} and Cu contents (r=0.78, and 0.89 respectively) (Table5).

Al, Fe, Mn, Co and Cr values do not show any significant correlation with the C_{org} content (Table 5).

3.4.2.2 Selective extraction analysis

Sequential exraction analysis were performed to determined the anthropogenic and /or natural inputs on metal distributions in the bay surface sediments. Metal contents of the geochemical phases were given in Table 6.

The highest values of Al, Fe, Zn, Co, and Cr varied between 2.2 % with 10.9 %, 3.8 % with 5.4 %, 18 % with 98 %, 4 % with 9 %, and 12 % with 51 % in the residual phase, respectively. In contrast, the highest values of Cu and Mn ranged from 6 % to 26 % in organic phase and from 32 % to 276 % in the Fe-Mn oxyhydroxide phase, respectively. While Fe and Cr values were generally lower than the detection limit of the methods (<0.05 and 0.08 µgL-1) in the exchangeable and carbonate phases, Al contents were also detected in the organic and residual (lithogenous) phases. Zn and Mn showed the highest values in Fe-Mn-oxyhydroxide phase, but Cu those in the organic phase along the bay. In addition, Cu, Zn, Mn and Co levels were relatively high in all geochemical phases.

Element	Exchangable phase	Carbonate phase	Fe-Mn-oxyhydroxide phase	Organic phase	Residual phase
Cu (ppm)	0.3-1.1	0.3-1	1.3-4.5	6-26	4-14
Zn (ppm)	0.1-2.3	0.8-37	15-121	14-46	18-98
Fe (%)	<0.05	<0.05	0.1-0.6	0.5-1.1	3.8-5.4
Mn (ppm)	1-13	6-51	32-276	32-241	32-176
Co (ppm)	0.1-1.3	0.1-2.2	0.3-3.7	0.2-9	4-9
Cr (ppm)	<0.08-4.5	<0.08	1.4-24	2-23	12-51
Al (%)	<0.03	<0.03	<0.03	0.1-0.4	2.2-10.9

Table 6. Metal distributions in different geochemical phases (%).

4. Discussion

DO concentrations of the water column were low in August 1999, after the earthquake, compare to that of other sampling periods. The low DO content was determined in all the stations of İzmit Bay, and particularly in the lower layer waters of the eastern and the central basins, being lower than the detection limit of the method (0.03 mgl^{-1}) (Figure 7b). The negative DO–SDO value along the water column suggested that the oxygen utilization was resulted from the decomposition of organic matter (Figure 5). The limited air-water exchange of free oxygen caused by the spreading petroleum from the refinery fire to the sea surface might be one of the main reason for lowering of DO content in water column. The highest oil concentration was determined in surface water of south of the central basin as 179.2 mgl^{-1} in August 1999 (Güven et al., 2000, Ünlü et al., 2000). The oil concentrations of the surface water are more than 500 µgl^{-1} in almost half of the western and central basins after the earthquake. In spite of high oil pollution levels of the surface water, the oil concentrations in the lower layer are between 13-55 µgl^{-1}in the Bay exception of north of the central basin in August 1999. This oil pollution level decreased to 10.5 mgl^{-1} in September 1999 and 3.3 mgl^{-1} in October 1999. The upper layer flows westward to Marmara Sea, while the lower layer flows into the Bay transporting oxygenated Mediterranean originated Marmara Sea waters in September and October 1999 (Güven et al., 2000). This current system provided the removal of the petroleum layer at the sea surface from İzmit Bay to the Marmara Sea and consequently DO concentrations increased in the water column accompanied by phytoplankton bloom (Figure 4a). Phytoplankton bloom was intense in the eastern basin (2,553,000 cell/l, Güven et al., 2000) and possibly the reason for the saturated DO content in this part of the Bay (Figure 5). Since the domestic and industrial wastewater system has been damaged by the earthquake, the nutrient input into the Bay increased, causing the extreme phytoplankton bloom (Okay et al., 2001). In spite of high DO concentrations of the upper layer, DHS is found in the lower layer of the eastern and the central basins (Figure 3). This striking condition clearly indicates the excess organic load that rapidly depositing at the bottom of

The Effect of Marmara (Izmit) Earthquake on the Chemical Oceanography and Mangan Enrichment in the Lower
Layer Water of Izmit Bay, Turkey

23

the Bay. Although no DHS data is available for the previous sampling period (August and September 1999), the establishment of this anoxic condition at the bottom might have started to develop earlier than October 1999. Earlier studies related to the ocenographic features of the Bay have never determined anoxic conditions in the water column (Morkoç, et al., 1988; Tuğrul et al., 1989; Morkoç et al., 1996).

The Marmara Sea water flows as the upper layer into the Bay in December 1999 and the current system is towards the interior of the Bay, whereas the lower layer flows out of the Bay (Güven et al., 2000). The available DHS formation in the eastern and the central basins is reduced or completely disappears in this month by the outflow of the lower layer (Table 2). This current system becomes reversed in February 2000, entering the lower layer and out-flowing the upper layer. The significant increase of DO concentrations of the upper layer in February 2000 might possibly indicate the replenishment of water column in İzmit Bay with oxygenated waters. This is in agreement with the vertical and spatial distribution of DO concentrations in February (Figures 3 and 4b). The thickness of the upper layer increases to 25-30 m suggesting the entrance of waters into the Bay. DO content of the both the upper and the lower layer slightly decreases in May 2000, together with increasing alkalinity (Figures 4b, 7b, 8 and 9). The reducing DO content in this month might be related with the water influx enriched with nutrients into İzmit Bay from the Black Sea (via the Marmara Sea) that receives increasing amount of freshwater inflow during spring (Oğuz and Sur, 1986; Tuğrul and Polat, 1995). In August 2000, DO concentration of the water column is significantly reduced (Figures 4 and 6), suggesting the enhanced consumption of DO by decomposition of high organic materials that possibly from the subsequent death of blooming phytoplanktons. In the eastern basin, the lowest pH is found in this month, supporting the increasing decomposition processes and the formation of DHS (Figure 9).

The formation of DHS leading to anoxia at the lower layer of İzmit Bay occurs in the eastern basin where the depths are shallower than 30 m and also locally in the deep site of the central basin where circulation is restricted (Table 2). After the Earthquake, in the central and the eastern basins, the formation of DHS is resulted from the spreading petroleum from the refinery fire to the surface waters and accumulation of high amounts of organic load from the damaged wastewater systems, and resuspension of bottom sediments releasing the DHS in the anoxic part of the sediment column. This is in agreement with the low DO concentrations of the water column in İzmit Bay during August and September 1999 (Figures 4a, 7a). The removal of anoxia at the bottom of the eastern and the central basins occurred in December 1999 by the replacing of water layers with the oxygenated Marmara Sea waters. DHS exists in the lower layer consistently throughout the sampling period in station 17 (Table 2), however its thickness varies. The reduced bottom current velocities (Algan et al., 1999) and topographic restriction of this small depression might be the reasons for the presence of DHS, by preventing the circulation.

In August 2000, DHS forms again in the eastern basin in low concentrations (Table 2). This re-occurrence of DHS is related with the extreme phytoplankton bloom. A high amount of organic matter produced from their death consumes oxygen for decomposition in the sediment. High decomposition rates might have led the depletion of DO in the overlying water column and consequent formation of DHS. The seasonal circulation pattern and timing of blooms in İzmit Bay were not different than the present as indicated by the previous studies (Oğuz and Sur, 1986; Tuğrul et al., 1989; Morkoç et al., 1996). DO content has never been fallen below 0.5 mgl⁻¹, and no DHS has been detected in İzmit Bay. Therefore, the re-occurrence of DHS a year after the Earthquake might indicate that İzmit Bay has not been completely return to its regular chemical oceanography. This may be explained by the fact that the amount of organic and possible inorganic wastes into İzmit Bay must have been considerably high and/or must have continued to discharge after the Earthquake. Increasing nutrients, phytoplankton blooms, rapid sedimentation of death organisms and decomposition processes constituted a successive cycle in İzmit Bay and intensified by the Earthquake at 17th August 1999. However, decomposition processes within this cycle might not be completed within a year.

The highest pH values found (8.9) at the upper layer compare to other months in the eastern basin confirms the increasing biological activity in October 1999 (Figure 8). During the respiration of phytoplanktons, dissolved CO_2 content of water column increases and consequently CO_3^{-2} and HCO_3^- anions increase. Increasing carbonate causes enhancement of alkalinity. The pH values become 7.9 at the lower layer (Figure 9) where the anoxic conditions are developed (Figure 3) and indicate the decomposition of organic matter.

Total metal contents in the İzmit Bay sediments increase towards to eastern basin. The eastern basin receives the highest inputs compare to other basins of the Bay (Morkoç et al., 2001). Ergin et al., (1991) suggested that the surface sediments in İzmit Bay are uncontaminated by anthropogenic pollution. However Yaşar et al., (2001) investigated that the heavy metal concentrations are highest in the eastern and central basins. The western basin was found generally unpolluted with respect to heavy metals in this study, also.

Selective extraction studies indicate that the metals are mainly found in the lithogenous, Fe-Mn-oxvhydroxide and organic fractions (Table 6). The results show that the main source of high metal concentrations in the İzmit Bay sediments is of anthropogenic origin. The highest metal values in these fractions are found in eastern basin sediments similar to total metal distributions.

5. Conclusions

Izmit Bay have been polluted by increasing industrial activities and domestic discharges since early 1980. However this abrupt event caused short-time drastic changes in the water column. Earthquake at 17 August 1999 initiated a fast variation in the chemical oceanography of polluted Izmit Bay. This variation includes the consumption of DO and formation of DHS in the lower layer. The refinery fire and damaged municipal waste effluents caused the reduction of DO in water column by preventing the oxygen transfer from air/ water contact and increasing organic wastes, respectively, and as a result DHS was formed. The increasing wastewater into the Bay stimulated the phytoplankton blooms that causes locally saturated DO concentrations in the eastern basin, however anoxic conditions were prevailing in the lower layer during autumn 1999. The changing circulation pattern during winter provided replenishment of the water column in Izmit Bay and removal of DHS. However, DHS formation established again in August 2000.

The distribution of total metals (Fe, Pb, Cu, Zn, Co, Cr and Cd) in both the water-column and surface sediments showed the influences of terrestrial anthropogenic inputs in the bay. The Mn enrichment in the lower-layer water of the central and eastern basins

The Effect of Marmara (Izmit) Earthquake on the Chemical Oceanography and Mangan Enrichment in the Lower
Layer Water of Izmit Bay, Turkey

25

originated from the occurring anoxic conditions after the Marmara (Izmit) earthquake. Selective extraction studies indicated that the metals were mainly found in the lithogenous, Fe-Mn-oxvhydroxide and organic fractions. The results underlined that the main source of high metal levels in Izmit Bay sediments is of anthropogenic origin. These conclusions reached by the selective extraction studies were supported by the "total" metal distributions along the bay.

6. Acknowledgements

The Captain, crew, scientists and technicians on board RV *Arar* of Institute of Marine Sciences and Management of Istanbul University, for their help during the collection of water samples. This work was supported by the Turkish Ministry of Environment.

7. References

Algan, O., Altıok, H.. and Yüce, H., 1999. Seasonal Variation of Suspended Particulate Matter in Two-layered İzmit Bay, Turkey. Estuarine, Coastal and Shelf Science 49, 235-250.

Altıok, H., Legovich, T. and Kurter, A., 1996. A case study of circulation and mixing processes in two-layered water system: İzmit Bay. Eight International Biennial Conference on Physics of Estuaries and Coastal Seas, Extended Abstracts, 8-12 September 1996, Netherlands Centre for Coastal Research, pp. 92-96.

Ayas, Z., & Kolankaya, D., 1996. Accumulation of some heavy metals in various environments and organisims at Göksu Delta, Türkiye, 1991-1993. Bulletin of Environmental Contamination and Toxicology, 56, 65-72.

Baştürk, Ö., Tuğrul, S., Sunay, M., Balkaş, T., Morkoç, E., Okay, O.S. and Bozyap, A., 1985. Determination of oceanographic characteristics and assimilation capacity of İzmit Bay. NATO TU-WATERS Project, Technical Report. Kocaeli, Turkey. TÜBİTAK-MRC Publication.

Beşiktepe, Ş.T., Sur, H.İ., Özsoy, E., Latif, M.A., Oğuz, T. and Ünlüata, Ü., 1994. The circulation and hydrography of the Marmara Sea. Progress in Oceanography 34: 285-334.

Bruland, K. W., Franks, R. P., Knauer, G.A., Martin, J. H., 1979. Sampling and analytical methods for the determination of copper, cadmium, zinc and nickel at the nanogram per liter level in seawater. *Anal. Chim. Acta* 105, 233-245.

Bruland, K. W., Coale, K. H., Mart, L., 1985. Analysis of seawater for dissolved cadmium, copper and lead: an inter comparison of voltametric and atomic adsorption methods. *Mar. Chem.* 17, 285-300.

DAMOC, 1971. Master plan and feasibility report for water supply and sewerage for Istanbul region. Prepared by the DAMOC Consortium for WHO, Los Angles, CA, Vol. III, Part II and III.

Doğan, E., Sur, H.İ., Güven, K. C., Polat, Ç., Kıratlı, N., Yüksek, A., Uysal, A., Algan, O., Balkıs, N., Ünlü, S., Altıok, H., Gazioğlu, C., Taş, S., Aslan, A., Yılmaz, N. and Cebeci, M., 2000. Su Kalitesi İzleme Çalışması. Technical Report. İSKİ. Deniz Bilimleri ve İşletmeciliği Enstitüsü, Istanbul Üniversitesi. (in Turkish).

Ergin, M., Saysam, C., Baştürk, Ö., Erdem, E. and Yörük, R., 1991. Heavy metal concentrations in surface sediments from the two coastal inlets (Golden Horn Estuary and İzmit Bay) of the northeastern Sea of Marmara. Chemical Geology, 91: 269-285.

Förstner, U. & Witmann, G.T.W., 1981. Metal pollution in the environment (p. 486). Berlin Heidelberg New York: Springer.

Greenberg, A.G., Trussel R.R., Clesceri, L.S., Franson, M.A.H., editors. Standard methods for the examination of water and wastewater, American Water Work Association (APHA, AWWA and WPCF). 16 th ed. Washington, 1985.

Güven, K.C., Sur, H.İ., Okuş, E., Yüksek, A., Uysal, A., Balkıs, N., Kıratlı, N., Ünlü, S., Altıok, H., Taş, S., Aslan, A., Yılmaz, N., Müftüoğlu, A. E., Gazioğlu, C., and Cebeci, M. 2000. İzmit Körfezi'nin Oşinografisi. 17 Ağustos 1999 Depremi sonrası İzmit Körfezi'nde Ölçme ve İzleme Programı. Technical Report. T. C. Çevre Bakanlığı. Deniz Bilimleri ve İşletmeciliği Enstitüsü, Istanbul Üniversitesi. (in Turkish).

Han, X.F., Banin, A., Su, Y., Monts, L.D., Plodinec, J.M. & Kingery, L. W., 2002. Industrial age anthropogenic inputs of heavy metals into the pedosphere. Naturwissenschaften, 89, 497-504.

Hu, H., 2000. Exposure to metals. Prim. Care, 27: 983-996.

Krauskopf, K. B., 1979. Intoduction to Geochemistry, (617 pp). Tokyo: McGraw-Hill Kogakusha.

Loring, D.H., & Rantala, R.T.T., 1992. Manuel for the geochemical analyses of marine sediments and suspended particulate matter. Earth-Science Reviews, 32, 235-283.

Lovley, D.R., Pihillips, E.J.P., 1988. Novel mode of microbial energy metabolism: organic carbon oxidation coupled to dissimilatory reduction of iron and manganese. Applied "Environmental Microbiology 54, 1472-1480.

Morkoç, E., Tuğrul, S. and Okay, S.O., 1988. Determination of Limiting Nutrients by Using Algal Bioassay Technique. Wastewater treatment and disposal studies. NATO-TU-WATERS, First Annual Report. TÜBİTAK-MRC Publications, Kocaeli, Turkey.

Morkoç, E., Okay, S.O. and Geveci, A.1996. Towards a Clean İzmit Bay. Technical Report. TÜBİTAK-MRC Publications, Kocaeli, Turkey.

Morkoç, E., Okay, S.O., Tolun, L., Tüfekçi, V., Tüfekçi, H., and Legoviç, T., 2001. Towarsd a clean İzmit Bay. Environmental International 26: 157-161.

Nealson, K.H., 1982. Microbiological oxidation and reduction of Iron: Holland, H.D., Schidlowski, M. (Eds.), Mineral Deposits and Evolution of the Biosphere. Dahlem Konferenzen, Springer, Berlin, pp. 51-56.

Nealson, K.H., Myers, C.R., 1990. Iron reduction by bacteria: potential role in the genesis of banded iron formation. American Journal of Science 290A, 34-45.

Nealson, K.H., Myers, Saffarani, D., 1994. Iron and manganese in anaerobic respiration: environmental significance, physiology and regulation. Annual Review in Microbiology 48, 31-343.

Oğuz, T. and Sur, H.İ., 1986. A numerical modelling study of circulation in the Bay of İzmit: Final Report. TÜBİTAK-MRC, Chemistry Department Publication, Kocaeli, Turkey, No. 187. 97 pp.

The Effect of Marmara (Izmit) Earthquake on the Chemical Oceanography and Mangan Enrichment in the Lower
Layer Water of Izmit Bay, Turkey

27

Okay, S.O., Legoviç, T., Tüfekçi, V., Egesel, L. and Morkoç, E., 1996. Environmental impact of land-based pollutants on İzmit Bay (Turkey): short-term algal bioassays and simulation of toxicity distribution in the marine environment. Archives of Environmental Contamination and Toxicology 31, 459-465.

Okay, S.O., Tolun, L., Telli-Karakoç, F., Tüfekçi, V., Tüfekçi, H.. and Morkoç, E., 2001. İzmit Bay (Turkey) Ecosystem after Marmara Earthquake and Subsequent Refinery Fire: the Long-term Data. Marine Pollution Bulletin, 42, 361-369.

Orhon, D., Gönenç, E., Tünay, O. and Akkaya, M., 1984. The prevention and removal of water pollution in the İzmit Bay : determination of technological aspects. Technical Report. İstanbul, Turkey: İTU-Civil Eng. Publ., 1984.

Sunda, W.G., Huntsman, S.A., 1995. Cobalt and zinc interreplacement in marine phytoplankton: biological and geochemical implications. Limnol. Oceanogr. 40, 1404-1417.

Strickland, J.D.H. and Parsons, T.R., 1972. A Practical handbook of seawater analysis 2nd. Ed., Oxford press, Ottawa, Bull. Fish. Bd. Can.

Pempcowiac, J., Sikora, A., Biernacka, E., 1999. Specification of heavy metals in marine sediments, their bioaccumulation by musssels. Chemosphere, 39: 313-321.

Tessier, A., Campbell, P. G. C., & Bisson, M., 1979. Sequental extraction procedure for the speciation of particulate trace metals. *Analyt. Chem.*, 51, 844-850.

Tuğrul, S., Sunay, M., Baştürk, Ö. and Balkaş, T.I., 1986. The İzmit Bay case study. In: G. Kullenberg (Ed.), The Role of Oceans as a Waste Disposal Option., pp. 243-275. Reidel, Dordrecht.

Tuğrul,S., Morkoç, E. and Okay, S.O., 1989. The Determination of Oceanographic Characteristics and Assimilation Capacity of the İzmit Bay. Wastewater treatment and disposal studies. NATO TU-WATERS, Technical Report. TÜBİTAK- MRC, Kocaeli-Turkey.

Tuğrul, S. and Morkoç, E., 1990. Transport and water quality modeling in the Bay of İzmit. NATO TU-WATERS Project, Technical Report. TÜBİTAK-MRC Publ. Kocaeli, Turkey.

Tuğrul, S. and Polat, Ç., 1995. Quantitative comparison of the influxes of the nutrients and organic carbon into the Sea of Marmara both from anthropogenic sources and from the Black Sea. Water Science and Technology, 32: 115-121 pp.

Ünlüata, Ü., and Özsoy, E., 1986. Oceanography of the Turkish Straits-First Annual Report, Volume II, Health of the Turkish Straits, I. Oxygen Deficiency of the Sea of Marmara, Institute of Marine Sciences, METU, Erdemli, İçel, Turkey, 81 pp.

Ünlüata, Ü., Oğuz, T., Latif, M.A. and Özsoy, E., 1990. On the Physical Ocenography of the Turkish Straits. In: Pratt, L. J., (Ed.), The Physical Oceanography of Sea Straits. Kluwer Academic Publishers, Netherland, pp. 25-60.

Ünlü, S., Güven, K. C., Okuş, E., Doğan, E. and Gezgin, T. (2000). Oil Spill Tüpraş Refinery Following Earthquake occured in 17 Aug 1999. Second International Conference, Oil Spills in the Mediterranean and Black Sea Regions, pp. 1-11, İstanbul, Turkey.

Yaşar, D., Aksu, A.E., Uslu, O., 2001. Anthropogenic Pollution in İzmit Bay: Heavy Metal Concentrations in Surface Sediments. Turk. J. Engin. Environ. Sci., 25, 299-313.

2

Earthquake Observation by Social Sensors

Takeshi Sakaki and Yutaka Matsuo
The University of Tokyo
Japan

1. Introduction

Many studies have examined observation and detection of earthquakes using physical sensors. These systems require highly accurate physical sensors located over a broad area, necessitating great expense to set up the supporting infrastructure.

Social media have garnered much attention recently and the number of social media users has been increasing. Social media are kinds of media for social interaction among users. Users create contents for themselves and exchange them on social media. Social media include many kinds of forms, including weblog, wikis, videos and microblogs. One of the biggest characteristics of social media is *user-generated contents*.

Social media users often make posts about what happened around them: live performance, sports events and natural disaster, including earthquake. Figure 1 depicts the graph of tweet counts and the sizes of earthquake on March 11th 2011, the day of the Great Eastern Japan Earthquake. It is apparent that tweet counts and earthquake occurrences are correlated. It means that when earthquakes occurs, social media users make posts about those earthquakes.

Along with the popularization of social media, new methods for earthquake observation are appearing. These method use information about earthquakes posted on the internet by users. For example, the web site *Did You Feel It?*, operated by the United States Geological Survey (USGS), gathers earthquake information from web-site users through a

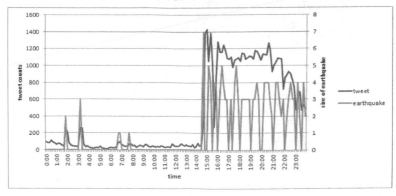

Fig. 1. Size of earthquakes and change of tweet counts on the day of the Great Eastern Japan Earthquake

questionnaire format(Intensity, 2005). From the Twitter web-site, *Toretter* extracts tweets that refer to earthquakes and estimates the location of an earthquake's epicenter using location information included with those tweets(Sakaki et al., 2010)

These methods treat social media users as sensors. We designate these virtual sensors as *social sensors*, which entail no costs. Unfortunately, such sensors provide a signal that is extremely noisy because users sometimes misunderstand phenomena, sleep, and are not near a computer.

We introduce these methods and explain a process for earthquake detection by analyzing social sensor information. We introduce current studies and services for earthquake observation using *social sensors* . Moreover, we explain *Toretter* as an example and describe its mechanisms.

2. Overview of earthquake observation by social sensors

We explain the basic idea of *social sensors* and introduce internet service users as social sensors to observe earthquakes.

2.1 Earthquake observation services performed by social sensors

We introduce four earthquake observation services that use information from internet users. In this chapter, we examine Toretter as an example. We explain its detailed mechanisms in the next chapter.

Did You Feel It?

The web site *Did You Feel It?*, which is operated by United States Geological Survey (USGS), is shown in Fig. 2. Through the internet, it gathers earthquake information from users who experienced those earthquakes directly (Intensity, 2005).

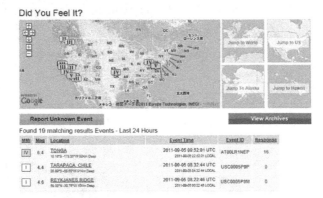

Fig. 2. Screenshot of *Did You Feel It?*

TED

The USGS also manages the Twitter Earthquake Detector (TED), which gathers tweets referring to earthquake occurrences from Twitter. They acquire location information and photographs attached to tweets and show this information related to maps(Survey, 2009).

iShake

The iShake project has developed a smartphone application (Fig. 3) that uses a phone to measure acceleration during an earthquake and report those data to researchers for processing (CITRIS, 2011). This project, conducted by UC Berkeley, is designed to create a system that moves beyond *Did You Feel It?*. Data from smartphone applications can complement data obtained from ground monitoring instruments, thereby improving the resolution and accuracy of earthquake intensity maps.

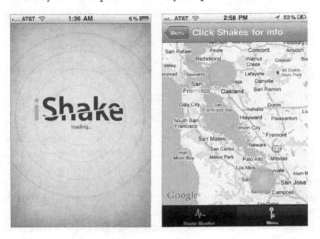

Fig. 3. Screenshot of *iShake*.

Toretter

Toretter extracts tweets referring to earthquakes and estimates the location of the earthquake epicenter using location information of those tweets(Sakaki et al., 2010). A temporal model and spatial model for earthquake detection are defined by social sensors. Then methods are proposed to detect earthquakes and to estimate the location of an earthquake epicenter automatically.

The Toretter mechanism is shown in Fig. 4.

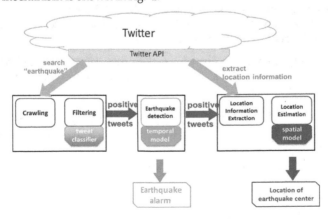

Fig. 4. Image of the Toretter mechanism.

First it collects tweets referring to earthquakes by crawling with the Twitter API and filtering the tweet messages using a tweet classifier. Second it tries to detect an earthquake from collected tweets based on a temporal model for earthquake detection. Finally, it extracts location information for each tweet from Twitter. The system uses that information and a particle filter to estimate the earthquake epicenter based on a spatial model for social sensors.

In this chapter, we explain methods of earthquake observation using social sensors according to the Toretter mechanism. We explain this entire process in the following section.

2.2 Overview of social sensors

We introduce the mode of *social sensors* and describe their features in comparison to physical sensors.

2.2.1 Basic idea of social sensors

Many methods and infrastructure can be used to observe events and natural phenomena using physical sensors: heavy traffic, air pollution, astronomical events, weather phenomena, and earthquakes are some examples. The basic mechanisms of such observations by physical sensors are presented on the right side of Fig. 5. First, a target event for observation occurs. Second, some sensors for the target event respond with a positive signal. Third, a central server collects signals from sensors and analyzes them. Finally, the server detects the target event or produces some observation values as output.

If users of social media observe an event, then similarly to physical sensors, they make posts about the event. For example, some Twitter users might post "Oh earthquake!" or "pouring rain, thunder & lightning " or "It's a double rainbow! & the moon is out. Beautiful!". These actions by users are analogous to the response of physical sensors to a stimulus: the users and sensors send a signal when an event occurs. Therefore, a user of social media is a sensor of a kind. We designate such sensors as social sensors.

An observation system incorporating social sensors is depicted on the left side of Fig. 5. First, an event occurs. Second, social media users make posts about the event. Third, the posts are collected at a central server and analyzed. Finally, the server detects the event or produces some observation value. This whole process corresponds to a process of observation by physical sensors, presented for comparison in Fig. 5

Methods for observing phenomena by physical sensors can be adapted to social sensors. Actually, some services based on social media use methods of observation resembling methods used with physical sensors.

Regarding Twitter users as social sensors, we can work with the following assumption.

1. Each Twitter user is regarded as a sensor. A sensor detects a target event and makes a report probabilistically.
2. Each tweet is associated with a time and location, which is a latitude–longitude pair.

2.2.2 Features of social sensors

Social sensors differ from physical sensors in some points. We describe features of social sensors in comparison to physical sensors.

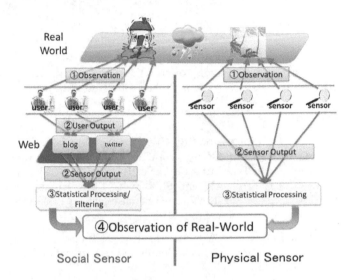

Fig. 5. Correspondence between event observation by social sensors and by physical sensors.

Social sensors are uncontrollable. They sometimes become inoperable because some users are not on-line; maybe they are sleeping or busy doing something else. They also function improperly more often than physical sensors because users misinterpret events more often than physical sensors. Therefore, it is necessary to know that social sensors are noisier than physical sensors and that their signals must be analyzed more carefully.

Social sensors, which are users of social media, are located over a wide area. They can give responses to events of many kinds, ranging from natural phenomena, such as earthquakes and hurricanes, to events related to human activities, such as heavy traffic, live performances, and elections. The extremely numerous social sensors all over the world present the possibility of responding to events of many kinds. In other words, detection of target events can be done with no cost to set up sensors. However, when using social media systems such as Twitter, which incorporate these social sensors, it is necessary to filter the signals (tweets) posted by social sensors (Twitter users) according to the event that is to be observed. Using some method, it is necessary to extract tweets referring to a target event. We summarize the features of social sensors and physical sensors in Table 1.

We explain these methods in the next section.

3. Tweet collection

In the first step portrayed in Fig. 4, it is necessary to collect tweets referring to an earthquake from Twitter. This process includes two steps: crawling tweets from Twitter and filtering out

features	physical	social
accuracy	high accuracy	noisy
versatility	target events only	event of any kind
cost	high	very low
processing	simple	complex

Table 1. Features of physical sensor and social sensors.

tweets that do not refer to the earthquake. For crawling and filtering tweets, we recommend using script programming languages, such as Python, PERL, and Ruby.

3.1 Crawling tweets from Twitter

To collect tweets or some user information from Twitter, one must use the Twitter Application Programmers Interface (API). Twitter API is a group of commands that are necessary to extract data from Twitter. Twitter has APIs of three kinds: Search API, REST API, and Streaming API. In this section, we introduce Search API and Streaming API, which are necessary to crawl tweets from Twitter. We explain REST API later because REST API is necessary to extract location information from Twitter information.

Additionally, it is known that Twitter API specifications are subject to change. When using Twitter API, it is necessary to know the latest details and requirements. They are obtainable from Twitter API documentation[1].

3.1.1 Twitter Search API

The Twitter Search API extracts tweets from Twitter, including search keywords or those fitting other retrieval conditions, in chronological order. It is possible to use language, date, location and other conditions as retrieval conditions.

When searching tweets including *earthquake* posted from 1 Aug 2011 to 5 Aug 2011, one might access the following URL:

http://search.twitter.com/search.json?q=earthquake&since=2011-09-01&until=2011-09-05

tweet time	user	tweet text
2011-09-04 04:47:10	user 1	The truth of 311 seismic terror.http://t.co/R9I6U9w 911 #earthquake #fukushima #japan #CNN #tsunami #prayforJapan
2011-09-04 04:47:09	user 2	FML! What did I say?! @..... RT @.... 24 HOUR EARTHQUAKE WARNING for San Diego, - 6.0+ likely - hey @....
2011-09-04 04:47:08	user 3	ML 2.3 SOUTHERN GREECE: Magnitude ML 2.3Region SOUTHERN GREECEDate time 2011-09-04 04:37:42.0 UTCLocation ...

Table 2. Search results of keyword *earthquake* after the conversion.

It is possible to obtain results in Fig. 6, as described in JavaScript Object Notation (JSON) format, which is a text-based open standard designed for human-readable data. It is possible to convert this result in Fig. 6 into Table 2 by parsing the result using a script programming language. Parameters that are often used to collect tweets are shown in Table 3 (This table is referred to Twitter API Documentation[2]).

---

[1] https://dev.twitter.com/docs
[2] https://dev.twitter.com/docs/api/1/get/search

```
- results: [
    - {
          created_at: "Sun, 04 Sep 2011 04:47:10 +0000",
          from_user: "911insidejob3",
          from_user_id: 261894525,
          from_user_id_str: "261894525",
          geo: null,
          id: 110212281127804930,
          id_str: "110212281127804928",
          iso_language_code: "en",
        - metadata: {
              result_type: "recent"
          },
          profile_image_url: http://a3.twimg.com/profile_images/1307316254/    normal.JPG,
          source: "&lt;a href=&quot;http://twittbot.net/&quot; rel=&quot;nofollow&quot;&
          gt;twittbot.net&lt;/a&gt;",
          text: "The truth of 311 seismic terror.http://t.co/R9I6U9w 911 #earthquake #fukushima
          #japan #CNN #tsunami #prayforJapan",
          to_user_id: null,
          to_user_id_str: null
      },
```

Fig. 6. Search results from Twitter Search API.

name	explanation	required	value
q	search keywords	required	-
rpp	the number of tweets to return per page	optional	up to 100
result type	search result of type	optional	mixed/recent/popular
until	tweets before the given date	optional	before today
since	tweets after the given date	optional	after 5 days ago
lang	restricts tweets to the given language	optional	jp/en/all/others

Table 3. Search conditions of Twitter Search API.

Some points must be considered when using Twitter Search API:

- It is possible to collect tweets posted only during the prior five days. It is not possible to search tweets posted six days ago.
- It is only possible to collect the latest 1500 tweets at one time.
  (Technically speaking, it is possible to access one page with a request and track pages back to the 15th page. One page includes 100 tweets at most. Therefore it is possible to acquire the latest 1500 tweets at one time.)
- One is limited to API requests.
  (No limit is published, but it is possible to access the Twitter Search API at least 500 times per hour.)

Therefore, we recommend the collection of tweets every 10 min or more often because it is impossible to crawl all tweets including *earthquake* if those tweets are posted at 2000 tweets per hour and one uses Twitter Search API every hour. Actually, tweets including *earthquake* were posted at more than 5000 per hour when the earthquake occurred on March 11, 2011.

Toretter requests the API command *search* 15 times every 5 min to collect the latest tweets each time: 180 command executions per hour.

3.1.2 Twitter Streaming API

The Twitter Streaming API extraction is defined in Twitter API documentation as follows:

The Twitter Streaming API enables high-throughput near-real-time access to various subsets of public and protected Twitter data.

Twitter Streaming API provides some methods shown in Table 4, of which *filter* method can be used to crawl tweets related to earthquakes.

command	explanation
filter	returns public statuses that match one or more filtering conditions.
firehose	returns all public statuses. A few companies have permission to access this command.
link	returns all statuses containing http: and https:.
retweet	returns all retweets
sample	returns a random sample of all public statuses.(ratio is about 1%)

Table 4. Streaming API methods.

Filter method returns public statuses that match one or more filtering conditions. All conditions of *filter* are presented in Table 5. It is possible to use the parameter *track* to collect tweets because keywords can be set as a condition value of *track*.

command	explanation
follow	returns public statuses that reference the given set of users.
track	returns public statuses that include specified keywords.
locations	returns public statuses that posted from a specific set of bounding boxes to track.

Table 5. Conditions of *filter* methods.

When using a *filter* command with the parameter keyword, *earthquake*, it is necessary to create a file called *tracking* that contains *track=earthquake*. Then one can access the following URL:

https://stream.twitter.com/1/statuses/filter.json

Streaming API also returns results in the form of JSON, shown in Fig. 6. Therefore, it is possible to parse those results in the same way as results obtained with Search API.

It is possible to collect tweets including *earthquake* in real time. Some points must be considered when using Twitter Streaming API:

• The prepared server must have sufficiently high specifications to process all data received from Twitter.

• It is impossible to use some characters in Twitter Streaming API
  (e.g., Japanese characters can not be used in Twitter Streaming API).

Using Toretter, we want to detect earthquakes in Japan. For that purpose, it is necessary to collect tweets including *earthquake* in Japanese. However, Japanese characters cannot be used in Twitter Streaming API. Therefore, Toretter uses the Twitter Search API to crawl tweets. To collect tweets of languages other than English, it is necessary to check whether that language is supported by the Twitter Streaming API.

tweet	real-time
SYF News: Magnitude 7.0 earthquake shakes Vanuatu; no tsunami alert	no
HOLY **** EARTHQUAKE	yes
Powerful earthquake rocks Vanuatu, no tsunami warnings (Newkerala)	no
AAAAAAAAAH earthquake !	yes
Holy ****, that earthquake scared the **** outta me	yes
a year on after our very first earthquake... and the shakes are still happening	no

Table 6. Sample tweets and relevance of real-time earthquake detection.

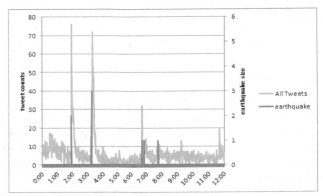

Fig. 7. Size of earthquakes and change of tweet counts on February 11, 2011

3.2 Filtering tweets using machine learning

We collected data from tweets including keywords related to earthquakes, such as *earthquake*, *shake*. Sample tweets are presented in Table6.

Those tweets include not only tweets that users posted immediately after they felt earthquakes, but also tweets that users posted shortly after they heard earthquake news, or perhaps they misinterpreted some sense of shaking from a large truck passing nearby. Figure 7 presents sizes of earthquakes and counts of Japanese tweets including the keyword *earthquake* on February 11, 2011. When the seismic activity reached its peak, the graph of tweets invariably showed a peak. However, when the graph of tweet counts showed a peak, the seismic activity did not necessarily show a peak. Some "false-positive" peaks of the graph of tweet counts arise from mistakes by users or some news related to earthquakes. Therefore, we must filter tweets to extract those posted immediately after the earthquake. We designate tweets posted by users who felt earthquakes as *positive* tweets, and other tweets as *negative* tweets.

Here, we describe the creation of a classifier to categorize crawled tweets into *positive* tweets and *negative* tweets, using Support Vector Machine: a supervised learning method.

3.2.1 Supervised learning

Supervised learning, a machine learning method, solves classification problem and regression problems analyzing training data. It is often used for spam mail filtering and gender estimation of Web users.

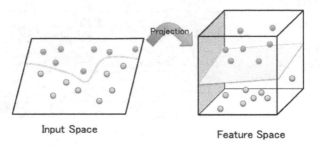

Fig. 8. Mechanism of Support Vector Machine.

Toretter uses Support Vector Machine (SVM), an extremely effective supervised learning method.

3.2.1.1 Support Vector Machine

SVM is a method used to create a classifier for two-class pattern classification. The SVM projects each training sample as points (as presented on the left side of Fig. 8) into multi-dimensional feature space. It creates a hyperplane that has the largest distance to the nearest training sample points of each class (as presented on the right side of Fig. 8). One must input positive samples and negative samples into SVM, which creates a classifier for two classes by searching the hyperplane.

To study them in detail, several books are useful (Bishop, 2006).

3.2.1.2 Process of creating a classifier using machine learning

Figure 9 depicts the process of supervised learning, which has three steps. We explain this process using an example of creation of a spam filter along the lines of Fig. 9

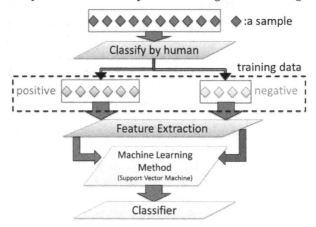

Fig. 9. Process of Machine Learning.

First, we prepare both sample collections of spam mails as positive samples and those of other mails as negative samples. Those must be classified manually by humans.

Second, we extract various features from samples. We must select effective features for classification. Effective features are those which positive samples seem to have and which negative samples do not seem to have, or vice versa. For example, all words included in samples are often used to create spam filters because we can infer that spam messages include words such as "Free!", "50% off!", and "Call now!" with high probability.

Third, we input both positive samples and negative samples with feature information and create a classifier for those samples. If inputting a new mail into the classifier, then it outputs a positive value or a negative value. If the output is positive, the new mail is regarded as a spam message.

3.2.2 Creation of sample data for the classifier

Positive samples and negative samples must be created manually. There are two points of consideration.

First, this process is very sensitive. One must classify positive tweets and negative tweets accurately. Therefore, it is necessary to acquire records of actual earthquakes. One must choose positive tweets referring to these earthquake records to classify them precisely.

Second, one must prepare equal numbers of positive tweets and negative tweets. The number of samples needed depends on the task. Generally, it is said that sample data must comprise 300–500 samples. Actually, one should increase the number of samples until finding the classification which provides sufficient performance.

3.2.3 Extraction of features from sample data

Next, one must select features of tweets for machine learning. In the spam mail filter example, words included in sample mails are chosen as features. Toretter uses features of three kinds. We explain them in detail and use the following sentence for explanation.

Oh! Earthquake happened right now!

Keyword features all words included in a tweet.
  example sentence → Oh, earthquake, happened , right, now
Statistical features number of words in a tweet message and the position of the search keyword within a tweet
  example sentence → number of words: *five*, the position of the search keyword: *second*
Context features words before and after a search keyword.
  example sentence → *Oh, happened*

Statistical features are the most effective in these three features according to results of our earlier research(Sakaki et al., 2010). It is guessed that this is true because people who came across an earthquake were surprised and in an emergency situation so that they tend to post short tweets such as "Oh! earthquake!" and "It's shaking".

Of course, these features can differ depending on language, country, and culture. Therefore, effective features should be chosen when creating a filter for tweets.

feature ID	feature	feature ID	feature
0	I	1	am
2	in	3	Japan
4	earthquake	5	right
6	now		
7	*number of words in tweets*		
8	*position of search keyword*		
9	*word before keywords* is Japan		
10	*word after keywords* is right		

Table 7. Sample features for SVM-Light.

3.2.4 Applying machine learning

Some machine learning methods can create a classifier for any problem: Naive Bayes classifier, Neural Networks, Decision Tree, and Support Vector Machine. In this chapter, Support Vector Machine is used for our explanation because it is said that SVM is a superior method for classification problems and regression problems, and many SVM software packages exist. We treat SVM-Light, which is a popular SVM tool, as an example in this chapter.

Creating a classifier demands three steps.

3.2.4.1 Create training data from tweets

First, it necessary to convert tweet data into a training data file format for SVM-Light. The training data file format of SVM-Light is

    <target> <feature>:<value> <feature>:<value> ... <feature>:<value> # <info>
    -1 1:0.43 3:0.12 9284:0.2 # abcdef

In this file format, each line corresponds to a single tweet. **<target>** expresses a polar of each tweet. +1 means positive and −1 means negative. **<feature>** expresses a feature ID of each feature and **<value>** expresses the weight of each feature in the tweet. Each feature should be assigned to each feature ID. For example, if one assigns each feature to each feature ID, as in Table 7, then a tweet conversion into a training data for SVM-Light as shown below.

    I am in Japan, earthquake right now → +1 0:1 1:1 2:1 3:1 4:1 5:1 6:1 7:7 8:5 9:1 10:1

You must run the following command to create a classifier for tweets after converting positive tweets collected into a training data file *training data file*.

    svm_learn "training data file" "model file"

svm_learn is a command in SVM-Light to create a model file for classifier. After running *svm_learn*, it is possible to obtain *model file* as an output of *svm_learn*. It is possible to classify the tweet command *svm_classify* with this model file. When classifying new tweets into a positive class and negative class, each tweet is converted into *test data* in the same format as *training data*. Then the following command is executed.

    svm_classify "test data file" "model file" "output file"

It is possible to obtain polars of each tweet in the *output file* New tweets are classifiable into a positive class and negative class by the classifier for tweets as described.

SVM-Light(Joachims, 2008), LIBSVM(Chih-Chung & Chih-Jen, 2011), and Classias(Okazaki, 2009) have compatibility such that the process we explain here is applicable to LIBSVM and Classias. (Toretter uses Classias for SVM tools.)

4. Earthquake detection from a time-series data using a probabilistic model

The second step of Fig. 4 detects an earthquake from positive tweets.

First, it is difficult to believe these tweets directly because some users misinterpret shaking caused by something other than an earthquake. Some ill-willed users post positive tweets to deceive others. This closely resembles physical sensors, and sometimes produces a wrong value. Therefore, we must process positive tweets to detect earthquakes with high accuracy, similarly to treating physical sensors.

Figure 10 depicts the sizes of earthquakes and counts of positive tweets filtered by SVM on Feb 11 2011. These two graphs are correlated: whenever an earthquake occurs, a peak appears in the graph of positive tweet counts. Therefore, we can detect earthquakes by detecting the peaks of positive tweet counts.

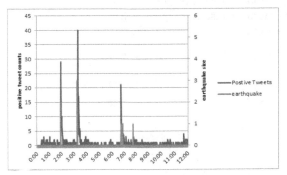

Fig. 10. Sizes of earthquakes and changes of filtered tweet counts Feb 11 2011.

Many methods have been used to detect peaks from time-series data for purposes such as burst detection(Kleinberg, 2002; Zhu & Shasha, 2003) and anomaly detection(Cheng et al., 2008; Krishnamurthy et al., 2003). Toretter uses a static rule *5 tweets in 5 min* that is calculated using an exponential function. We explain this method hereinafter.

4.1 Temporal model

To detect an earthquake using physical sensors, we must calculate the probability of earthquake occurrence based on signals from those sensors. Similarly, we must calculate the probability of earthquake occurrence from signals of social sensors. In this subsection, we explain the temporal model we use to calculate this probability.

Figure 11 presents graphs of positive tweet counts during earthquakes. In Fig. 11, the green line shows an exponential function. As shown here, the green line resembles the red line,

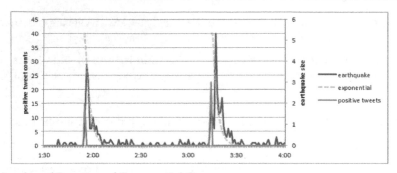

Fig. 11. Number of Tweets and Exponential Curve.

the graph of positive tweet counts. It can be inferred from these graphs that this frequency distribution of positive tweets is an exponential distribution, as expressed by the following equation(Sakaki et al., 2010).

$$f(t\lambda) = ke^{-\lambda t} \tag{1}$$

We express the number of sensors producing positive value at time t in $n(t)$. Here, $n(t)$ is equal to the number of positive tweets at time t. If n_0 sensors produce positive value at $t = 0$, then we can calculate the number of sensors for which the response is a positive value at time t using the following equation.

$$n(t) = n_0 \cdot e^{-\lambda t} \tag{2}$$

Therefore, we can calculate N_{t_a}, the number of sensors that produce a positive value from time 0 to time t_a, as presented below.

$$N_{t_a} = \sum_{t=0}^{t_a} n(t)$$
$$= n_0 \sum_{t=0}^{t_a} e^{-\lambda t}$$
$$= n_0 \frac{1 - e^{-\lambda(t_a+1)}}{1 - e^{-\lambda}} \tag{3}$$

We define the false-positive ratio of a sensor as p_f. In this case, we assume that we have n sensors and that all n sensors have the same false-positive ratio equally. The probability of all n sensors producing a false alarm is p_f^n. Therefore, the probability of earthquake occurrence can be estimated as

$$P(n) = 1 - p_f^n. \tag{4}$$

From Eq. 3, Eq. 4, we can calculate the probability of earthquake occurrence at time t_a.

$$p_{occur}(t) = 1 - p_f^{N_{t_a}}$$
$$= 1 - p_f^{n_0\left(1 - e^{-\lambda(t_a+1)}\right)/\left(1 - e^{-\lambda}\right)} \tag{5}$$

4.2 Setup the condition for detection trigger

In the Toretter system, we detect an earthquake when *five positive tweets arrive in 5 min*, which means *five sensors produce positive signals in 5 min*. In this subsection, we explain how to determine this condition.

We set $\lambda = 0.34, p_f = 0.35$ (taken from our earlier research) to Equation (5) , by which we can calculate the probability of earthquake occurrence. When obtaining n_0 positive tweets, and given that we would like to make an alarm with false-positive ratio less than 1%, we can calculate t_{wait} as

$$t_{wait} = -\frac{1}{0.34} log \left(1 - \frac{1.264}{n_0} \right) - 1. \tag{6}$$

If we set $t_{wait} = 5$, then we can calculate $n_0 = 4.1$ from Eq. 6. Therefore, the trigger for earthquake detection is set as *five positive tweets come in 5 min* in Toretter. The trigger used for detection of earthquake calculation can be determined using an exponential function, as described in this section.

5. Location estimation from tweets

In this section, we explain a means to estimate the location of an earthquake epicenter by analyzing tweets. First, we introduce the kinds of location information to be acquired from tweets. Next, we explain methods to estimate the location of the earthquake epicenter.

5.1 Extracting location information from tweets

Two kinds of information are applicable for location estimation from tweets: using location information in the Twitter user profile or using *geotag* attached to tweets.

5.1.1 Location information in user profiles

The twitter user profile includes the location information of users. Of course, not all users make their location information public on the internet, but a sufficient number of users do so (This number varies among countries.).

For earthquake detection, we collect positive tweets. We extract the location information of users who post those positive tweets for earthquake epicenter location estimation. Twitter REST API must be used to extract location information of users from Twitter.

Twitter REST API is one Twitter API included among all methods to use basic functions of Twitter. Many methods of using REST API exist. We use the *users/show* method to obtain user information. To extract user information of Twitter user *TwitterAPI*, it is necessary to access the following URL.

    http://api.twitter.com/1/users/show.json?screen_name=TwitterAPI
    &include_entities=true

It is possible to obtain results in Fig. 12, which is described in JSON format, in the same manner as that used for Twitter Search API. It is possible to know from the result in Fig. 12 that Twitter user TwitterAPI resides in *San Francisco, CA*.

Some points to consider when using Twitter REST API are the following:

```
{
    profile_sidebar_fill_color: "a9d9f1",
    protected: false,
    id_str: "6253282",
    notifications: false,
    profile_background_tile: false,
    screen_name: "twitterapi",
    name: "Twitter API",
    display_url: null,
    listed_count: 9143,
    location: "San Francisco, CA",
    expanded_url: null,
    show_all_inline_media: false,
    contributors_enabled: true,
    following: false,
    geo_enabled: true,
    utc_offset: -28800,
    profile_link_color: "0094C2",
    description: "The Real Twitter API. I tweet about API changes, service
    about Twitter and our API. Don't get an answer? It's on my website.",
    profile_sidebar_border_color: "0094C2",
    url: http://dev.twitter.com,
    time_zone: "Pacific Time (US & Canada)",
```

Fig. 12. User information extraction from Twitter Search API.

- Some users do not register their location information, or register non-location data, such as *in a dream, anywhere*. Such non-location data should be ignored.
- API requests are limited.
  (The limit is published: it is possible to access the Twitter Search API about 150 times per hour without authorization.)

It is possible to access REST API 150 times per hour. This limit is sufficient to extract user information for location estimation of an earthquake epicenter because the earthquake-related tweets posted in the 5 min after an earthquake are most often fewer than 100. To expand the limit, one must register with Twitter and obtain an authorization called OAuth, according to the Twitter API Documentation[3].

Moreover one must convert location information acquired from Twitter into a latitude–longitude pair because human beings can understand places expressed by the names of places, such as *San Francisco*, but a computer can not understand where that place is. One must treat location information in the format of a latitude–longitude coordinate pair. At present, some web services can convert geographical names into a latitude–longitude coordinate pairs, such as the Google Maps API and Yahoo Maps API. Here we explain the Google Maps API.

To convert *San Francisco* into a a latitude–longitude coordinate pair, one can access the following URL.

> http://maps.google.com/maps/api/geocode/json?address=San
> %20Francisco&sensor=false&language=en

Results are obtainable as in Fig. 13, which is described in JSON format, in the same manner as Twitter API. It is possible to convert *San Francisco* into *latitude* $= 37.7749295$, *longitude* $= -122.4194155$.

Location information related to an earthquake can be acquired as described above.

---

[3] https://dev.twitter.com/docs/auth

```
[
  - results: [
    - [
      + address_components: [ ··· ],
        formatted_address: "San Francisco, CA, USA",
      - geometry: [
        + bounds: [ ··· ],
        - location: [
            lat: 37.7749295,
            lng: -122.4194155
          ],
          location_type: "APPROXIMATE",
        + viewport: [ ··· ]
        ],
      + types: [ ··· ]
      }
    ],
    status: "OK"
}
```

Fig. 13. Result of geographical name converted using Google Maps API.

5.1.2 Geotags attached to each tweet

Some tweets have an attached geotag, which includes a latitude–longitude pair acquired from GPS. If positive tweets related to an earthquake include tweets with attached geotags, then it is possible to use these geotag data for location estimation. Geotag data can be extracted using the Twitter Search API. Therefore, GPS data can be obtained if stored when using crawl for those tweets by the Twitter Search API.

Geotag data are more accurate than location information of the Twitter user profile because they are acquired from GPS. Nevertheless, it is unusual that positive tweets referring to an earthquake include a sufficient number of tweets with attached geotags to estimate the earthquake epicenter location. Actually, a combination of location information of Twitter users and geotag should be used.

5.2 Location estimation using Bayesian filtering

If one can obtain sufficient location information from positive tweets, then estimating the location of the earthquake epicenter can be done using the information. Nevertheless, that information is often inaccurate. Alternatively if they are precise, then users might still be posting far from the earthquake epicenter. Therefore, it is preferred that the location of the earthquake epicenter be estimated probabilistically.

Several methods can be used to estimate the location of events from sensor readings using Bayesian Filters: Kalman filters, Multihypothesis tracking, Grid-based approaches, Topological approaches, and Particle filters.

We use particle filters as an example for explanation. Particle filters have high performance in belief, accuracy, robustness, and variety according to an evaluation by Fox et al. (Fox et al., 2003). Moreover particle filters work better to detect earthquakes from Twitter in the experiments by Sakaki et al. (Sakaki et al., 2010).

5.2.1 Spatial model

Each tweet is associated with a location. We describe a method that can estimate the location of an event from sensor readings. To define the problem of location estimation, we consider the evolution of the state sequence $\{x_t, t \in \mathbf{N}\}$ of a target, given

$$x_t = f_t(x_{t-1}, u_t), \quad f_t : \mathcal{R}_t^n \times \mathcal{R}_t^n \to \mathcal{R}_t^n,$$

where f_t is a possibly nonlinear function of the state x_{t-1}. Furthermore, u_t is an i.i.d. process noise sequence. The objective of tracking is to estimate x_t recursively from measurements, as

$$z_t = h_t(x_t, n_t), \quad h_t : \mathcal{R}_t^n \times \mathcal{R}_t^n \to \mathcal{R}_t^n,$$

where h_t is a possibly nonlinear function, and where n_t is an i.i.d. measurement noise sequence. From a Bayesian perspective, the tracking problem is to calculate, recursively, some degree of belief in the state x_t at time t, given data z_t up to time t.

Presuming that $p(x_{t-1}|z_{t-1})$ is available, the prediction stage uses the following equation.

$$p(x_t|z_{t-1}) = \int p(x_t|x_{t-1})p(x_{t-1}|z_{t-1})dx_{t-1}$$

Here we use a Markov process of order one. Therefore, we can assume that

$$p(x_t|x_{t-1}, z_{t-1}) = p(x_t|x_{t-1}).$$

In the update stage, Bayes' rule is applied as

$$p(x_t|z_t) = p(z_t|x_t)p(x_t|z_{t-1})/p(z_t|z_{t-1}),$$

where the normalizing constant is

$$p(z_t|z_{t-1}) = \int p(z_t|x_t)p(x_t|z_{t-1})dx_t.$$

To solve the problem, several methods of Bayesian filters are proposed such as Kalman filters, multi-hypothesis tracking, grid-based and topological approaches, and particle filters. For this study, we use particle filters, both of which are widely used in location estimation.

Additionally, we must consider the nonuniform distribution of Twitter users when we apply Bayesian filters to *social sensors* because *social sensors* are arranged non-uniformly to a greater degree than normal physical sensors are.

5.2.2 Location estimation using a particle filter

A particle filter is a Bayes filter that approximates a state probabilistically. It is a sequential Monte Carlo method. For location estimation, we maintain a probability distribution for the location estimation at time t, designated as the belief $Bel(x_t) = \{x_t^i, w_t^i\}, i = 1 \ldots n$. Each x_t^i is a discrete hypothesis related to the location of the object. The w_t^i are non-negative weights, called *importance factors*, which sum to one.

The Sequential Importance Sampling (SIS) algorithm is a Monte Carlo method that forms the basis for particle filters. The SIS algorithm consists of recursive propagation of the weights and support points as each measurement is received sequentially.

We use a more advanced algorithm with re-sampling. We use weight distribution $D_w(x, y)$, which is obtained from the Twitter user distribution to assess the biases of user locations[4] . The algorithm is shown as follows:

1. **Initialization**: Calculate the weight distribution $D_w(x, y)$ from Twitter users' geographic distribution in Japan.
2. **Generation**: Generate and weight a particle set, which means the N discrete hypothesis.
   (a) Generate a particle set

   $$S_0 = (s_0^0, s_0^1, s_0^2, \ldots, s_0^{N-1})$$

   and allocate them evenly on the map, as

   $$particle \; s_0^k = (x_0^k, y_0^k, w_0^k)$$

   $x, longitude; y, latitude; w, weight$
   (b) Weight them based on weight distribution $D_w(x, y)$.
3. **Re-sampling**
   (a) Re-sample N particles from a particle set S_t using weights of respective particles and allocate them on the map. We allow re-sampling of more than that of the same particles.
   (b) Generate a new particle set S_{t+1} and weight them based on weight distribution $D_w(x, y)$.
4. **Prediction**: Predict the next state of a particle set S_t from Newton's motion equation.

$$(x_t^k, y_t^k) = (x_{t-1}^k + v_{x_{t-1}} \Delta t + \frac{a_{x_{t-1}}}{2} \Delta t^2,$$

$$y_{t-1}^k + v_{y_{t-1}} \Delta t + \frac{a_{y_{t-1}}}{2} \Delta t^2)$$

$$(v_{x_t}, v_{y_t}) = (v_{x_{t-1}} + a_{x_{t-1}}, v_{y_{t-1}}, a_{y_{t-1}})$$

$$a_{x_t} = \mathcal{N}(0; \sigma^2), \quad a_{y_t} = \mathcal{N}(0; \sigma^2).$$

5. **Weighing**: Re-calculate the weight of S_t by measurement $m(m_x, m_y)$ as follows.

$$dx_t^k = m_x - x_t^k, \quad dy_t^k = m_y - y_t^k$$

$$w_t^k = D_w(x_t^k, y_t^k) \cdot \frac{1}{(\sqrt{2\pi}\sigma)}$$

$$\cdot exp\left(-\frac{(dx_t^{k2} + dy_t^{k2})}{2\sigma^2}\right)$$

6. **Measurement**: Calculate the current object location $o(x_t, y_t)$ by the average of $s(x_t, y_t) \in S_t$.
7. **Iteration**: Iterate Steps 3, 4, 5, and 6 until convergence.

---

[4] We sample tweets associated with locations and obtain a user distribution that is proportional to the number of tweets in each region.

6. Evaluation and application

In this section, we explain how to evaluate results of experiments and describe points that should be considered when applying these methods.

6.1 Selection of the target area

Three conditions must be met to apply methods for earthquake observation from social media.

The first is that a sufficient number of people use Twitter in a targeted area. The second one is that several earthquakes occur each year for a target area. The third one is that infrastructure should be set up in a target area.

These three conditions are needed in each step of earthquake detection and location estimation. A sufficient number of tweets and a certain number of earthquakes are needed to create a classifier for tweets and to estimate the locations of earthquake epicenters. Accurate logs of earthquakes are also necessary to calculate the false-alarm probability of social sensors and to evaluate the earthquake detection system performance.

If creating a classifier and setting a trigger for earthquake detection in an area and applying them in another area, then the third condition is not indispensable. However, the first condition and the second condition are necessary in both areas.

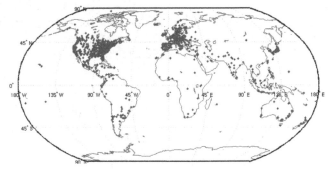

Fig. 14. Twitter user map.

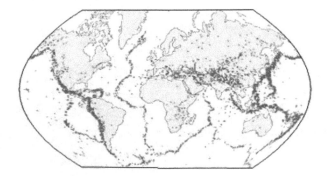

Fig. 15. Earthquake map.

Figure 14 depicts the Twitter user distribution map and Fig. 15 depicts an earthquake occurrence distribution map. Earthquake detection using information from Twitter users is applicable in overlapping areas of these two maps: for example, Japan, the west coast of the U.S., Indonesia, Turkey, Iran, and Italy.

The number of Twitter users has been increasing continuously. Therefore, those areas can probably be expanded. Additionally, if one uses social media other than Twitter, then overlapping areas might be changed.

Therefore, a target area should be chosen very carefully to apply the methods described in this chapter.

6.2 Evaluation of earthquake detection

To evaluate the performance of earthquake detection and earthquake epicenter location estimation, one must collect earthquake data from some organizations. Those data must include information about an approximate time point of an earthquake and approximate position of an earthquake epicenter. Moreover, it is better that they include the exact time of an earthquake, the longitude and latitude of an earthquake epicenter, and the seismic intensity of earthquakes in each region.

For example, the Japan Meteorology Agency (JMA) publishes an earthquake database on the Web, which includes a time, magnitude, and earthquake intensities at each point of area, a place of earthquake epicenter of all earthquakes above level 1 on the Japanese seismic intensity scale[5]. The USGS publishes similar data on the Web[6].

Data of such kinds can be obtained by crawling. They can be used to create training data for classifiers and to evaluate the performance an earthquake detection system.

7. Conclusion

Our research is an early approach to using Twitter as a social sensor for earthquake observations. It is meaningful that we apply methods by ordinary physical sensors to earthquake detection by social sensors. Furthermore, we present the possibility of earthquake detection without installing numerous physical sensors. The method is effective for earthquake observations in some countries where a few seismic sensors exist. However, it is difficult to detect earthquakes occurring in oceanic areas or less populated areas using methods we introduced in this chapter. Therefore, we must verify that earthquake detection by social sensors is effective when we apply these methods. Furthermore, the applicable scope of the earthquake observation by social sensors can be extended considering a stochastic gradient, more detailed probabilistic models, and so on. Many subjects remain to be explored in future work.

8. References

Bishop, C. M. (2006). *Pattern Recognition and Machine Learning*, Vol. 4 of *Information science and statistics*, Springer.

---

[5] http://www.seisvol.kishou.go.jp/eq/shindo_db/shindo_index.html
[6] http://neic.usgs.gov/neis/qed/

Cheng, H., Tan, P.-N., Potter, C. & Klooster, S. (2008). Data mining for visual exploration and detection of ecosystem disturbances, *Proceedings of the 16th ACM SIGSPATIAL international conference on Advances in geographic information systems - GIS '08* p. 1.

Chih-Chung, C. & Chih-Jen, L. (2011). LIBSVM-a Library for Support Vector Machine. URL: *http://www.csie.ntu.edu.tw/ cjlin/libsvm/*

CITRIS (2011). iShake -Mobile Phones as Seismic Sensors-. URL: *http://ishakeberkeley.appspot.com/*

Fox, D., Hightower, J., Schulz, D. & Borriello, G. (2003). Bayesian filtering for location estimation, *IEEE Pervasive Computing* 2(3): 24–33.

Intensity, M. (2005). Did You Feel It ? Citizens Contribute to Earthquake Science, *Technical Report March*, U.S. Geological Survey. URL: *http://earthquake.usgs.gov/earthquakes/dyfi/*

Joachims, T. (2008). SVM-Light. URL: *http://svmlight.joachims.org/*

Kleinberg, J. (2002). *Bursty and hierarchical structure in streams*, ACM Press, New York, New York, USA.

Krishnamurthy, B., Sen, S., Zhang, Y. & Chen, Y. (2003). Sketch-based change detection: methods, evaluation, and applications, *Proceedings of the 3rd ACM SIGCOMM Conference on Internet Measurement*, ACM, pp. 234–247.

Okazaki, N. (2009). Classias. URL: *http://www.chokkan.org/software/classias/index.html.en*

Sakaki, T., Okazaki, M. & Matsuo, Y. (2010). Earthquake shakes Twitter users: real-time event detection by social sensors, *Proceedings of the 19th international conference on World wide web* pp. 851–860.

Survey, U. S. G. (2009). Twitter Earthquake Detector (TED). URL: *http://recovery.doi.gov/press/us-geological-survey-twitter-earthquake-detector-ted/*

Zhu, Y. & Shasha, D. (2003). Efficient elastic burst detection in data streams, *Proceedings of the ninth ACM SIGKDD international conference on Knowledge discovery and data mining - KDD '03*, ACM Press, New York, New York, USA, p. 336.

3

Earth Observation for Earthquake Disaster Monitoring and Assessment

Huadong Guo, Liangyun Liu, Xiangtao Fan, Xinwu Li and Lu Zhang
Key Laboratory of Digital Earth Science, Center for Earth Observation and Digital Earth,
Chinese Academy of Sciences, Beijing
China

1. Introduction

China is a country where earthquakes and many other disasters happen often. After earthquakes, roads are damaged, traffic is blocked off, secondary disasters occur frequently, weather conditions become adverse, and communications are interrupted, which makes it difficult to gather data from stricken regions. And the big problem for recovery operations is that there is no accurate information about the situation. Earth observation technology, which has many advantages including high-speeds, maneuverability, and macro- to micro-level observation, has shown its importance for gathering information about stricken regions and making reasonable recovery decisions.

Optical Earth observation technology can provide vivid images for target interpretation and disaster information extraction. Maneuverable, flexibile airborne optical observation technology can especially provide real-time surface images, which also obtains information about collapsed houses, broken roads, geological disasters, barrier lakes and so on. It plays an important part in disaster mitigation activities (Guo et al., 2010a). Synthetic aperture radar (SAR) not only has the capability of all-weather monitoring, but also is sensitive to geometric shape and movement, which becomes an efficient tool to analyze and evaluate recent earthquakes (Guo et al., 2000; Guo et al., 2010b). Multi-mode SAR data can provide many kinds of information for disaster research. Wide-mode SAR images and In-SAR images are important methods for detecting terrain deformation. Wide-mode SAR images can analyze the faulted zone and lithologic characteristics in stricken regions from a macro-level, because it acquires large-scale image (Guo et al., 2000). In-SAR images yield information about surface deformation size and spatial distribution acquired from two-scene repeat-pass data (Massonnet & Feigl, 1998). Polarimetric SAR images, due to the sensitivity to target structures, can be used to extract the distribution of collapsed buildings.

After the Wenchuan and Yushu earthquakes, some departments took full advantage of airborne and satellite remote sensing technology, or unmanned aerial vehicles, to obtain images of the disaster area, which played a very important role in disaster emergency monitoring and disaster assessment and reconstruction (Guo et al., 2010ab; Singh et al., 2010; Liou et al., 2010). Besides monitoring targets directly affected by the disaster, such as collapsed buildings (Lei 2009), remote sensing can observe secondary damage such as barrier lakes, collapse, landslide, debris flow et al. (Cui et al, 2008; Wang et al, 2008; Liu et al., 2009; Huang et al., 2009; Ge et al., 2009; Xu et al., 2009; Han et al., 2009; Zhuang et al., 2010; Zhang et al., 2010; Xu et al., 2010).

The Chinese Academy of Sciences (CAS) immediately arranged a cooperative data acquisition program of airborne and satellite remote sensing data after Wenchuan and Yushu earthquakes and obtained 17 categories of more than 500 scenes of satellite images and high-resolution optical and microwave airborne remote sensing data. 8.7 TB of high-resolution data were freely provided initially to 16 ministries and 28 units, and an additional 3.5 TB were later downloaded from the network. At the same time, a study on remote sensing monitoring methods for post-earthquake secondary geological disasters was carried out, which played an important role in the disaster response. This paper focuses on three aspects, including optical Earth observation technology for monitoring secondary geological disasters, multi-mode radar Earth observation for post-earthquake deformation analysis, and an earthquake disaster simulation evaluation system using the results of seismic disaster remote sensing.

2. Detecting geological disasters using optical technology for Earth observation

Optical technology for Earth observation can provide visual images for disaster target interpretation and disaster information extraction. Airborne optical technology is one of the main instruments for Earth observation, with its mobility and flexibility to provide real-time disaster remote sensing and surface images. With disaster mitigation work done to remotely sense secondary disasters after the Wenchuan earthquake, including barrier lake breaches, road damage, and landslides and debris flows, we analyze and discuss technical methods and applications of optical technology for Earth observation in monitoring secondary geological disasters.

2.1 Extracting background information from the disaster area in Wenchuan

The Wenchuan earthquake occurred in the Longmen mountain fault zone. Longmen mountain runs in the general northeast to southeast direction, about 500km from Guangyuan to Ya'an. Longmen mountain is one of China's typical nappe structures, a tectonic rock sheet along a imbricated thrust to the basin mainly formed in the Mesozoic and early Cenozoic (Wang et al, 2001). The Wenchuan earthquake occurred in the crustal brittle-ductile transition zone, and was a shallow earthquake with a focal depth of 10 km to 20 km and longer duration, so its destructiveness was huge (Bi et al, 2008).

After the Wenchuan earthquakes, CAS urgently arranged airborne and satellite data coordinate acquisition plans and obtained 41 scenes of post-disaster, high-resolution satellite data, and 105 scenes of pre-disaster and concurrent high-resolution archive satellite data. An optical remote sensing airplane carrying an advanced ADS40 aviation camera obtained high-resolution (0.5 - 0.8 m) optical pictures of the disaster area totaling 5 TB with a coverage area of 23,000 km².

2.2 Remote sensing monitoring and analysis of barrier lakes after the Wenchuan earthquake

High-resolution ADS40 optical images of the disaster area were used to analyze the barrier lake for the first time. In the coverage area, 51 barrier lakes were detected, some with a bead-like distribution. The location, area, water level and height, and area of the dam body were detected according to a monitoring algorithm of barrier lake risk factors and 1:50,000 DEM data. The risk conditions, geology, and distribution of the 51 major barrier lakes were evaluated to support urgent relief work. The research indicated that the distribution of barrier lakes and spatial features of the earthquake fracture zone were identical.

2.2.1 Barrier lake volume detection algorithm

The water level and area of the barrier lakes were first estimated using high-resolution airborne images. The capacity of the barrier lake was then calculated using elevation contours, and the calculation was based on DEM data with a resolution of 25 m, which were interpolated from a 1:50,000-scale topographical map.

The method to calculate the reservoir capacity involves the following steps:

The water surface area was derived from high-resolution airborne images. Then the water surface elevation (h_s) was acquired by overlapping the water surface with the DEM data, since the elevation of the water surface is a constant. If there is some small shift between the orthorectified ADS40 image and the DEM data, the interpreted water surface should be adjusted slightly to ensure all interpreted water surfaces' borderline is located at the same altitude. Meanwhile, the elevation of the midline of the river (h_r) was directly recorded from the 1:50,000-scale topographic map. Therefore, the elevation difference and the water level could be calculated.

The capacity of the barrier lake was calculated by an integral approach. The capacity V at the water elevation of H is:

$$V(H) = \sum_{i=1}^{n} S_i \times \Delta h \qquad (1)$$

where Δh is the integration interval, n is the equally parted cells number of the elevation drop from the water surface elevation (h_s) to the elevation of the midline of the river, (h_r), $\Delta h = (h_s - h_r)/n$ is the integration interval of each cell, and S_i is the water surface area at the elevation of $h_r - (i - 1)\Delta h$, which can be automatically derived from the DEM data. The capacity and area of all the 51 barrier lakes were calculated by this method. According to their capacities, the barrier lakes were clustered into three types: Type I (large-sized) with a capacity over 3,000,000 m³; Type II (medium-sized) with a capacity between 1,000,000 and 3,000,000 m³; and Type III (small-sized) with a capacity less than 1,000,000 m³

2.2.2 Risk assessment of the barrier lakes

Barrier lakes formed in an earthquake will result in extreme flooding when they burst. Therefore, the risk assessment of barrier lakes becomes very important. The dimensionless blockage index (DBI) was introduced by Casagli and Ermini (Ermini et al, 2003; Liu et al., 2009) to evaluate the stability of a dam:

$$DBI = log(\frac{A_B \times H_d}{V_d}) \qquad (2)$$

where V_d is the volume of the dam, which is the dominant parameter of stability since it determines the gravity of the dam; A_b is the area of the basin, which is the primary parameter of instability since it determines the runoff in the basin; and H_d is the height of the dam, which is an important parameter for evaluating the stability of the barrier lake when confronted with overflow. The smaller the DBI value, the more stable the barrier lake. It is difficult to calculate the dam volume with the remote sensing image without in-situ measurement. An approximate estimation of dam volume is to multiply the dam area with its height, and thus Eq. (2) can be written as:

$$DBI \approx \log\left(\frac{A_b \times H_d}{H_d \times S_d}\right) = \log\left(\frac{A_b}{S_d}\right) \tag{3}$$

where S_d is the dam area.

2.2.3 Detection results of quake lakes in the Wenchuan earthquake

We interpreted the barrier lake surface and the dam area from the high spatial resolution ADS40 airborne images and located the position of the landslide forming the barrier lake. The basin area of the barrier lake was then extracted with the DEM and hydrological data. Therefore, the reservoir capacity and DBI of the barrier lake could be calculated according to Eqs. (1) and (3). The monitoring information of barrier lakes within airborne remote sensing data coverage shows that:

i.   Generally, the slope of the landslided are steep, and most of them are over 20 degrees. This condition causes the formation of barrier lakes.

ii.   DBI values can reliably reflect the stability of the barrier lakes. A lower DBI value indicates a more stable barrier lake, but the risk of a secondary disaster is higher if breaches and overflows occur. According to the overflow time of the two barrier lakes located at Xiaojia Bridge, Anxian County, and Tangjiashan, Beichuan County, the barrier lakes in the Wenchuan earthquake could survive for more than 30 days if the DBI were smaller than 4.0.

iii. The Wenchuan earthquake generated 10 large-sized and 14 medium-sized barrier lakes. Therefore, immediate attention should be paid to those barrier lakes with serious and continuous breaches.

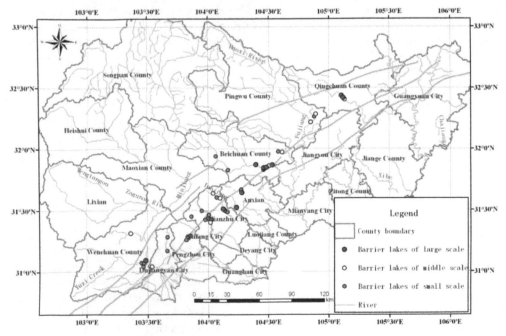

Fig. 1. Distribution map of barrier lakes caused by the Wenchuan earthquake (Liu et al., 2009)

2.2.4 Geological conditions and spatial distribution of the barrier lakes

The Wenchuan earthquake triggered many geological hazards, including collapses and landslides along river valleys. Some of the large masses of land fell into the river valleys and formed a number of barrier lakes. Figure 1 shows the distribution of 51 barrier lakes through the interpretation of remote sensing images. From the distribution map, the barrier lakes were apparently along the Yingxiu-Beichuan fault, and the distribution was consistent with the direction of the fault zone. There were a series of high-risk barrier lakes distributed along the rivers such as the Jianjiang River's upstream in Beichuan County, the Mianyuan River's upstream in Mianzhu City, and the Pingshui River in Shifang City.

2.3 Remote sensing monitoring and analysis of roads damaged by Wenchuan earthquake

High-resolution ADS40 optical airborne remote sensing images and other data were used to analyze and locate some national and provincial highways in seriously damaged areas. The process included analyzing qualitatively, orientatively and quantitatively different factors and classes of blocked and damaged roads from landslip, debris flows, river bank collapse, barrier lakes, earthquake disruption and ground fissuring. These monitoring and analysis results may give transportation department powerful information support.

2.3.1 Road blockage and damage conditions in badly stricken areas

The main remote sensing road blockage and damage condition detection focuses on national and provincial highways. There are 5 national and provincial highways in the badly stricken

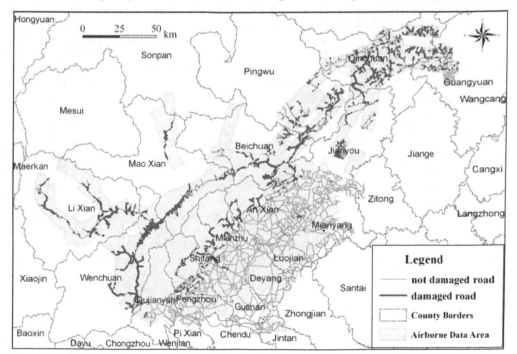

Fig. 2. Map of damaged roads after the Wenchuan earthquake

area, with an overall length of 573.82 km. This includes 3 national roads with a total length of 394.07 km: National Highway 213 (239.563 km) from Xuankou Town, Wenchuan County to Minjiang Village, Mao County, National Highway 212 (31.814 km) from Baolun town, Guangyuan City to Lijia Ping, Qingchuan County, National Highway 317 (122.692 km) from Siboguo Village, Li County to Wenchuan County. It also includes 2 provincial highways with a total length of 179.75 km: Provincial Road 105 (50.85 km) from Anchang Town, An County to Beichuan County and Provincial Road 302 (128.90 km) from Mao County, Jiangyou City.

The Wenchuan earthquake caused more than five national and provincial road blockages and damage, including 808 points in total with a total length of 170.17 km, which accounted for 29.66% of the total length of paths in the badly stricken area, in which, National Road 213 was most seriously damaged, then National Road 317 and Provincial Road 105. Damage to the others was comparatively light. The conditions of damaged roads is described in Figure 2.

2.3.2 Damage level and distribution condition of hard-hit areas
The blocked roads were obviously segmented and the worst parts appeared to have a cluster distribution. Outside of the observed areas, the roads were light damaged and without major disasters such as landslips.

2.3.3 Category, scale and causation of damaged paths
According to remote sensing monitoring and analysis, the blockage paths of the worst-damaged places were caused by geological disasters such as landslips, falling debris, mud-rock flows and ground fissures. The distribution of these disasters was related to the break structure, drape structure, and rock broken under stress.

Fracture tectonic belt was the most important role in road damage. The magnitude-8 earthquake happened on the fault zone of Longmen Mountain, which is composed of many approximately parallel disruptions oriented towards the northeast with a length of more than 500 km and 50 km in width. The Yingxiu-Beichuan fault and Wenchuan-Maoxian fault were the most important parts of the Longmen Mountain fault zone, and the most serious damage to National Road 213 was along these two fault zones.

The background of the badly stricken area's geological structure is very complex with a long evolutionary history. The rocks in hard-hit areas were squeezed extremely because of the long-term activities of well-developed faults and fold faults. These broken rocks provide a large amount of material for potential falling, mud-rock flows (with fragment flows) and landslips. This material slipped rapidly under the force of gravity and piled up on the lower roads and caused blockage in the harder-hit area.

In addition, the earthquake's power is the reason that roads in harder-hit area were blocked and damaged. When the earthquake occurred, there were both southeast-direction thrust extrusion and dextral shear in the Longmen Mountain Fault Zone (Chen Yuntai etc., 2008), which made steep mountains, high-angle clockwise slope and extremely broken rock strata lose their balance and dependence. Then under the action of gravity, broken rock rapidly slid downward, or collapsed and accumulated in low-lying areas on the highway, which damaged and blocked the road.

2.4 Secondary geological disaster analysis of Wenchuan earthquake
Using high-resolution ADS40 optical aviation remote sensing images and comprehensive, conventional data, we evaluated secondary geological disasters, such as collapses,

landslides, and debris flows. Using investigations by remote sensing and field survey results combined with the Ministry of Land and Resources, PRC emergency investigation data, a secondary geological hazards information acquisition model was established. It allowed for the monitoring and rapid extraction of secondary geological disaster data. Researchers studied the characteristics and distribution of secondary geological disasters, completed the interpretation of 11 heavily-damaged area's geologic disasters and their geological background, and analyzed the geological hazards' distribution, intensity, scale and distribution regularity. The study revealed that the development of secondary geological disasters had obvious clustering. In the area, seismic geological hazards were along the Longmen Mountain Fault Zone and mainly caused by the Beichuan-Yingxiuwan fracture control.

2.4.1 Remote sensing information extraction method for secondary geological disasters

The special weather conditions in Wenchuan make the ratio of vegetation coverage higher. The study on vegetation destruction in this area shows the severity of geological disasters and supports the job of disaster evaluation and post-disaster reconstruction.

First, remote sensing data were acquired from before and after the earthquake. Extracting the information map of vegetation was done with standardized resolution and registration. Second, high spatial resolution images were used to check causes of vegetation change including geological disasters, human destruction, agricultural changes and so on. The third step was to remove the vegetation change information caused by non-geological disasters and make a classification according to remote sensing image identification of secondary geological disasters. Last, statistics were calculated and analysis conducted to make a difference map.

i.   Standardize resolution and make registration

In order to reduce the extraction difference of vegetation caused by different spatial resolutions, we standardized the spatial resolution of the images by two-dimensional cubic convolution before the vegetation information extraction. Then map-to-map control points were adopted and registered by a polynomial method. Because the registration method uses vegetation information difference extraction before and after the disaster, the registration precision should be controlled within 1 pixel.

ii.   extract vegetation information

According to the special vegetation spectrum, we can efficiently extract the vegetation information from pre- and post-disaster images. The formula is as follows:

$$NDVI = \frac{R_{NIR} - R_R}{R_{NIR} + R_R} \tag{4}$$

where NDVI is vegetation index, R_{NIR} is reflectivity of near infrared wave, and R_R is reflectivity of infrared wave. Comparing the results with a real vegetation coverage map to fix the threshold, we can extract the vegetation coverage map and change it into a vector diagram.

iii.   extracting the vegetation information difference from geological disasters

Based on the extraction of a vegetation information vector graph, the difference between the two figures can be extracted by doing differential operation. The high spatial resolution data can be used to correct elements caused by the difference and then reject the difference in vegetation information caused by non-geological factors. Finally, according to remote sensing image identification of secondary geological disasters, they can be classified.

2.4.2 Secondary geological disaster remote sensing monitoring results

The overall results of monitoring secondary geological disaster are shown in Figure 3, where the red stands for landslides, blue for debris flows, and green for collapses. As can be seen from the figure, the landslide occurred mainly in Beichuan, Pingwu and Qingchuan counties; debris flow occurred mainly in Wenchuan, Maoxian and Li counties; and collapse occurred mainly in Wenchuan and Beichuan counties. According to statistics, the total disaster area is about 29,000 square kilometers, and the secondary geological disaster area is 2,250 square kilometers (7.8% of the total).

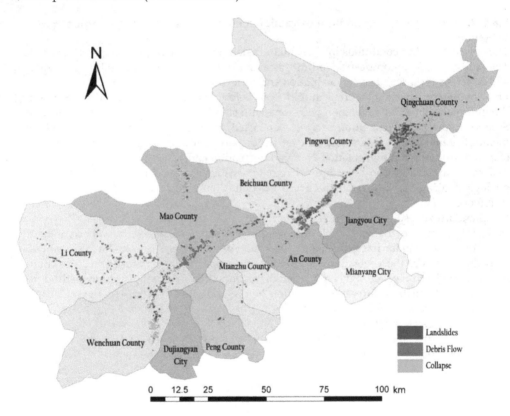

Fig. 3. Map of secondary geological disasters occurring in the heavily-hit areas of the Wenchuan earthquake

3. Multi-mode radar imaging technology for monitoring earthquake deformation

Synthetic aperture radar (SAR) multi-mode data can provide a variety of types of information for earthquake research. We take Yushu earthquake, which occurred in Yushu

County, Qinghai Province, China, on April 14, 2010, as the research object and use RADARSAT-2 and ALOS-PALSAR repeat-pass SAR interferometry data to analyze earthquake deformation characteristics.

3.1 Yushu earthquake area background and data acquisition

The Yushu earthquake occurred in the Garzê-Yushu Fault Zone. The fault strike runs in a northwest direction for a length of nearly 500 km, and has a fracture width from 50 to several hundred meters. From analysis of the plate tectonics, it can be concluded that the source of this earthquake was in the Qinghai-Tibetan Plateau, located in the north of the collision zone in the Himalayas, which was formed by the subduction of the Indian plate toward the Eurasian continent. Because of this plate subduction, lateral sliding of the internal blocks of the Qinghai-Tibetan Plateau occurred, which caused the northward shift of the plateau and its internal blocks and finally, the formation of strike-slip fault systems with different scales at the edge of the blocks. Zhang et al. (2010) inverted the moment tensor solution using wave-form data from global stations. From this solution and the background of the fault tectonics, it can be concluded that the fault with a trend of 119° and a dip of 83° was the earthquake rupture. The breaking process was determined based on teleseismic data from the 35 global stations. Two active regions on the fault surface were identified. One was located near the micro-epicenter, and the other was located to the southeast at a distance of 10 to 30 km. The latter had the greater slip, 2.4 m, and was a near vertical sinistral strike-slip fault.

The study uses SAR data including RADARSAT-2 wide-mode data and ALOS PALSAR repeat-pass data. The RADARSAT-2 wide-mode data was acquired on April 21, 2010, with a spatial resolution of 40 meters and an incident angle of 21 degrees. ALOS PALSAR data, including two pre-earthquake and post-earthquake scenes, were acquired on January 15, 2010, and April 17, 2010, respectively. Table 1 shows the PALSAR data parameters for repeat-pass SAR interferometry.

Sensor	Date	Orbit	Frame	Perpendicular baseline (m)	Temporal baseline (d)
PALSAR	2010-1-15	487	650	700.5	92
PALSAR	2010-4-17	487	650		

Table 1. PALSAR parameters for SAR interferometry

3.2 The method of extracting earthquake geological characteristics and surface deformation information from SAR data

Ground-fissuring phenomena are often a reflection of different lithological characteristics. SAR image brightness and texture structure can reflect the degree of fissuring. In addition, radar waves are sensitive to the linear structure (Guo, 1996, 1997), so using SAR imagery can help interpret tectonic information.

Interferometric SAR is an important means of extracting surface deformation because it can measure it precisely in three-dimensional space, including small deformations of the surface, and can achieve high spatial resolution observation of surface changes in large areas. Interferometric SAR images of the same area at different times by SAR sensor were obtained at different time SAR complex images. Then we process SAR images acquired at different times to obtain an interferogram. SAR interferograms show electromagnetic wave transmission path length variation from the SAR antenna to the target in two images. Electromagnetic wave transmission path length changes are generally subject to the following three factors: satellite position changes, surface changes, and atmospheric changes. Product by the satellite position changes is terrain interferometry phase, which produced by surface changes is the atmospheric phase. Generally speaking, the SAR interference phase can be expressed as type (5).

$$\Phi_{IFG} = W\left\{ \Phi_{topo} + \Phi_{defo} + \Phi_{atm} + \Phi_{noise} \right\} \tag{5}$$

where $W\{\}$ is a phase winding operator, and deformation interference phase in type (5) Φ_{defo} represents monitoring surface deformation ability of SAR interferometry. In order to obtain the deformation interference phase information, the atmosphere in the mathematical model of SAR interference measurement will be classified as noise signal, which directly considers the SAR interferometry phase as the terrain interferometry phase and deformation interference phase (Rosen, 2000), and then the removal of the terrain interferometry phase can obtain surface deformation information.

3.3 Multi-mode SAR data and the Yushu earthquake area evaluation results
3.3.1 Yushu earthquake area lithology and SAR image fraction analysis results
To further analyze the regional earthquake geology, wide-swath RADARSAT-2 SAR data were acquired on April 21, 2010, with HH polarization, a spatial resolution of 40 m, and an incident angle of 21°. Combined with geological data, the study area can be divided into A to E regions (Figure 4).Because radar waves are sensitive to the linear structure (Guo, 1996,

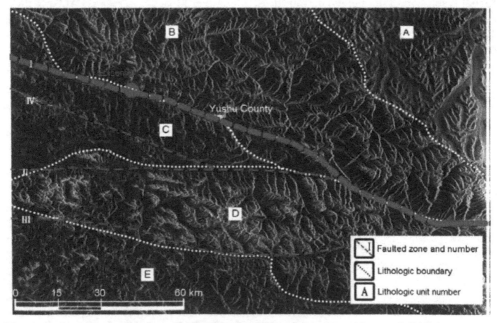

Fig. 4. Geological analysis from RADARSAT-2 HH polarization wide-swath SAR imagery. (from Guo et al., 2010b)

1997), based on the SAR image interpretation and existing research results of active tectonic plates (Deng, 2007), four main faults of this area have been interpreted as follows: the main faults I and IV are oriented in a northwest-southeast direction; fault IV developed in the limestone areas of the map; and faults II and III are distributed in an east-west direction. According to the structural composition of the faults and existing active tectonics results, the main fault I is a strike-slip sinistral fracture.

3.3.2 Yushu earthquake area InSAR deformation extraction analysis result

Using Doris InSAR data processing software and SRTM3 DEM data with 90 m resolution, the two-pass differential interferometry method was used to process the ALOS PALSAR data. We then get the seismic deformation interference phase image shown in Figure 5.

The radar interferogram clearly shows the spatial distribution of the surface deformation field caused by the Yushu earthquake. The coseismic deformation field within the image is about 82 km long and about 40 km wide along the fault. From the distribution of the interferometric fringes caused by the Yushu seismic deformation field, we can see that the distribution of the coseismic deformation is centered on the Garzê-Yushu fault zone, which is the triggered fault (Figure 4, the main fault I), and is parallel to this fault. From the distribution pattern of the interferometric fringes, we can see that the direction and density of the interferometric phase change are different for the two sides of the fault. From the southernmost point A to the fault direction, the interferometric fringe phase indicates an increasing trend from south to north. To the north of the fault, the interferometric fringe phase shows a decreasing trend from north to south. From the whole interferometric phase distribution, the change in the line of sight is left-lateral, revealing significant seismogenic fault sinistral strike-slip properties. It corresponds with the result of wide swath SAR image interpretation.

The seismogenic fault is in a northwest-southeast direction. Along the seismogenic fault zone, there are two major areas with large surface deformation, shown as ① in Figure 5(b) and ② in Figure 5(c). Position ① corresponds with the instrumental epicenter calculated by the National Earthquake Network, and ② corresponds with the macroscopic epicenter. From enlarged views of the interferogram of the instrumental epicenter area in Figure 5(b) and macroscopic epicenter regions in Figure 5(c), we can also see that the radar interferometric fringes change intensely around the instrumental epicenter, while the central region of the macroscopic epicenter has an apparent decorrelation due to the large surface deformation. Both of the two regional seismic fault slip dislocations are relatively large, but that of the latter region, which is close to the city of Yushu, will inevitably lead to stronger tremors for the city of Yushu and the surrounding area, where rupture has been the predominant cause of enormous casualties and economic losses.

PALSAR operates in the L-band, and a color change cycle in the interferogram represents 11.8 cm in the line of sight. According to the interferometric fringes analysis, on the north of the fault, the maximum sinking displacement in the line of sight is 11.8×3=35.4 cm. Since the surface near the epicenter was damaged during the earthquake, the coherence of the corresponding region in the two radar images is very low and cannot form effective interferometric fringes. Therefore, it is reasonable to conclude that one fringe remains on each side of the fault in the low coherence area, and the cross-fault displacement in the line of sight should not be less than 11.8×8=94.4 cm.

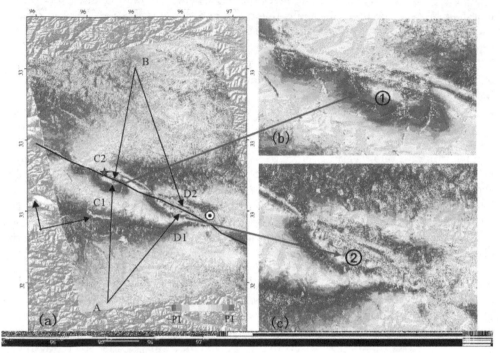

Fig. 5. Coseismic deformation map from ALOS PALSAR data, (a) Differential interferometric phase map; (b) Differential interferometric phase of instrumental epicenter; (c) differential interferometric phase of macro epicenter. A, B, C1, C2, D1 and D2 in (a) represent the different positions and in (b) and in (c) are two large deformation areas (from Guo et al., 2010b).

3.4 Extraction and analysis of collapsed buildings from polarimetric SAR

Large-scale earthquakes severely damage people's lives and property. Fast, accurate, and effective collapsed buildings a monitoring and evaluation after earthquake using remote sensing provides an important scientific basis and decision-making support for government emergency command and post-disaster reconstruction.

RADARSAT-2 polarimetric SAR data (FQ mode, ascending) on April 21, 2010 from Yushu County were used to extract the distribution of collapsed buildings. The resolution of the image is about 8 m and the incidence angle is 21°. From this polarimetric SAR data, the H-α–ρ method (Guo et al., 2010b) was used to extract the spatial distribution of building collapse caused by the earthquake in the Yushu urban area. At the same time, for comparison and analysis, a manual interpretation map obtained from the high-resolution airborne optical image was also collected and shown in Figure 6. From the collapsed buildings extraction result, the reasons for the severe earthquake damage to buildings are also discussed and assessed.

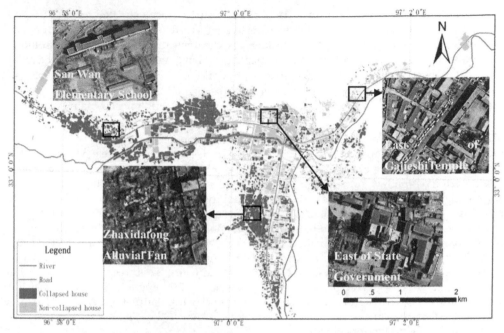

Fig. 6. Spatial distribution of collapsed buildings interpreted from airborne images and sample images for typical regions.

3.4.1 H–α–ρ method

From the RADARSAT-2 polarimetric data (FQ mode) and the polarimetric decomposition model, a new H–α–ρ method that uses only one post-earthquake SAR image was proposed to identify collapsed buildings. This method mainly utilizes three important polarimetric parameters to extract the collapsed buildings. These parameters are H, α and ρ, Where H is entropy, representing the random level of target scattering, α is the averaged scattering type and ρ is the circle polarization correlation coefficient which is very sensitive to artificial objects. H and α are obtained by using Cloude's H–α decomposition(Cloude, 1996,1997).The ρ of uncollapsed buildings is high while that of collapsed buildings is low. However, ρ is also related to the surface roughness. For a low roughness surface, the ρ is also high. Therefore, it is necessary to remove the disadvantaged influence of the bare soil surface before this parameter can be used for building identification(Ainsworth etc al.,2008). The circular polarization correlation coefficient ρ can be expressed as Eq. 6(F.Mattia, 1997),

$$\rho_{RRLL} = \frac{< S_{RR}S_{LL}^* >}{\sqrt{< |S_{RR}|^2 >< |S_{LL}|^2 >}} \qquad (6)$$

Where $S_{RR} = iS_{HV} + \frac{1}{2}(S_{HH} - S_{VV})$, $S_{LL} = iS_{HV} - \frac{1}{2}(S_{HH} - S_{VV})$. For the Yushu urban area, three main land cover types were categorized: collapsed buildings, uncollapsed buildings, and bare soil surface. The basic process using H–α–ρ method to identify the collapsed buildings is as follows: 1) the extraction of the bare soil surface. H and α are obtained using the $H-\alpha$ decomposition theorem, then, the bare soil surface was extracted with $H<0.5$ and $\alpha<42°$ (Cloude, 1996,1997). 2) Using statistical analysis, the ρ, which can discriminate between uncollapsed and collapsed buildings, is determined. From the high- resolution optical data, typical areas of collapsed and uncollapsed buildings are selected from the radar image to analyze the statistical characteristics of ρ, and the appropriate threshold value of ρ is obtained. 3) From the threshold value of ρ, the separation of collapsed and uncollapsed buildings is conducted, and the distribution map of collapsed buildings is obtained.

3.4.2 Result and analysis of collapsed building extraction

Figure 7 shows the collapsed building distribution of the Yushu urban area extracted by the $H-\alpha-\rho$ method. The rate of building collapse is about 58%. To verify the result, two test sites were selected as follows: (a) a severely damaged area, and (b) an almost undamaged area. Comparing the extraction results with the manual interpretation results from the airborne optical image with a resolution of 0.33 m shown as a' and b' in the lower left corner, we can see that the result of the collapsed buildings' extraction is consistent with the result from the optical manual interpretation. Furthermore, to verify the effectiveness of this method, the recognition rates of collapsed buildings and uncollapsed buildings are analyzed from two sample regions (more than 10000 pixels) from a and b test sites, respectively. From the statistical analysis, the recognition rate for collapsed buildings is 88% and that for uncollapsed buildings is 80%. It should be noted that the polarimetric SAR could play a more important role in the collapsed building extraction if the weather conditions were unsuitable for obtaining optical data.

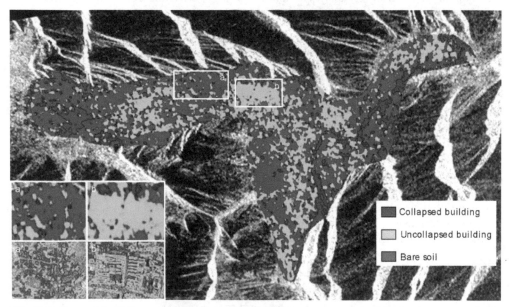

Fig. 7. Collapsed buildings distribution of Yushu County extracted with the $H-\alpha-\rho$ method from polarimetric SAR data, a and b in the lower left corner are the detailed images of the two test sites, a' and b' show the corresponding manual interpretation results from the airborne optical image(from Guo et al., 2010b).

From the collapsed building distribution, it is clear that the degree of collapse is related to the distance from the main fault. The buildings, which were constructed in the alluvial zone, had very poor earthquake resistance because their foundations were weak and most were built of earth and wood.

4. The foundation of an earthquake condition simulation and evaluation system

Based on remote sensing image two-dimensional spatial information systems, interpretation and analysis has been the main means of seismic disaster assessment. Although two-dimensional spatial information systems have the macroscopic and overall characteristics, it also has inevitable defects in the earthquake disaster assessment (Xiao et al, 2001, Li et al, 2007), such as in accurate expression of three-dimensional information, an inability to record non-uniform three-dimensional spatial entities, and a lack of a basis for uneven spatial entity description of lake water, landslides, collapsed buildings and so on. Therefore, it is necessary to establish an evaluation system based on three-dimensional spatial information of earthquake disaster information for analysis.

For the Wenchuan and Yushu earthquakes, using three-dimensional simulation and evaluation with advanced Earth observation technology, we established realistic 3D terrain model using the relatively sophisticated disaster assessment model and successfully created an earthquake disaster simulation and evaluation system. Based on multi-sensor, multi-temporal, and multi-resolution remote sensing images and 1: 50,000 scale DEM data, we produced technology for large 3D terrain modeling and interactive real-time rendering. The 3D simulation system provides a more intuitive disaster analysis method for major collapse, landslide, and debris flow disasters. Using red-blue 3D imaging techniques to get airborne remote sensing stereo images and reconstruct three-dimensional scenes, we can efficiently extract data on damage to housing and improve the disaster analysis accuracy using red and blue stereo glasses. In a 3D environment, not only can the geo-spatial relationship between objects be shown, but also the topological relations between spatial objects. Practice has proven that this kind of three-dimensional assessment is more efficient and reliable.

4.1 Three-dimensional terrain modeling and visualization

We use 1: 50,000 topographic vector data of the disaster area for error analysis and to eliminate gross errors of contour and control point data. We use a difference algorithm for vector contour data to obtain high-resolution DEM raster data. In order to rebuild 3D virtual scenes, we use a merging method of aerial remote sensing images and Landsat TM images to generate the terrain texture. In severe disaster areas, we acquired high-resolution data, TM images. Then combined with DEM data, the aerial remote sensing data can be corrected precisely. Finally, the merging of aviation data and TM data yield the an image of the entire area, which is then mapped to the three-dimensional terrain model to form a virtual 3D environment.

The complicated terrain of earthquake-stricken areas and large data of three-dimension model after overlaying images and DEM have brought challenges for real-time rendering. This paper proposes a multi-resolution triangular grid dynamic geographic model based on computing vision, established a simplified algorithm based on multi-resolution vision. Using a multi-resolution scene model algorithm, this paper resolves the real-time interactive roaming difficulty of large-scale three-dimensional terrain data. First, according to the amount of data and resolution of images, the magnitude n is established. According to the given format, a model file of different resolutions is generated, and then when modeling real-time rending, a series of irregular grids is used to imitate the terrain. According to their distance from the point of view and complexity of the terrain, we choose the relevant resolution terrain model within sight of the study area that is the closest to the view point. The more complex the terrain is, the higher the terrain series are, the more triangle grids in the drawing area, the more sophisticated the display terrain, and the higher the resolution. Conversely, if the view point is farther away or the terrain is flat, the series of the topography of the area shows will be lower, the number of mesh triangles will be fewer, the terrain rougher, and the resolution lower. Thus, minimizing the number of triangles and reducing memory consumption can make the images and models have identical effects or gaps in a given range, closest to the real terrain.

In this paper we take the improved adaptive quaternary tree to construct layers of details. The quaternary tree index and the grading mode of organization management for large-scale three-dimensional scenes can provide the chance for   real-time interactive roaming analysis. In order to ensure the smoothness of scene rendering, we adopt a pre-loaded cache before the scene into the visible range. Then based on real-time rendering of the scene's rectangular range and the quaternary tree index, we can access related scenes' serial numbers quickly, which can be pre-loaded into memory. The scheduling management strategy can improve the rendering efficiency significantly, achieve real time, interactive terrain rendering, and improve the efficiency of disaster evaluation and analysis in the 3D environment.

4.2 Three-dimensional terrain modeling and visualization

Exposed areas along the river valley include a regional north-east-trending thrust fault. The area exhibits many faults and tectonites, which are weak and vulnerable and form the detachment surfaces. Due to the earthquake's physical destruction combined with heavy rain, a large number of landslides, mud-rock flows, and other geological disasters occurred. Our system provides a qualitative and quantitative analysis and monitoring of these secondary disasters. A 3D system for the analysis of secondary disasters has the following advantages:

1. The three-dimensional system can reproduce the true disaster scene, and secondary disasters can be observed directly, including landslides and debris flows and their causes, occurrence and trends (Figure 8a).
2. Using DEM data, the analysis of landslide height and volume can give quantitative estimates of damage due to the scope and extent of landslides.
3. The visual angle can rotate in a 3D system; providing omnidirectional observation of the target, which has obvious advantages in making decisions about rescue routes and strategies.

One example is a statistical analysis of the secondary disaster in Chenjiaba. High-resolution aerial remote sensing images from May 28 were integrated into the system. The system gives information on landslides and mud-rock flows in the area from Beichuan County, Zhixin Village, to Pingwu County Yaogouli. The study identifies the Chenjiaba section landslide debris and other secondary geological disasters as the most serious. The system also marks the distribution of the disaster and builds disaster level categories. There is a total of 135 landslides, covering an area of 508 square kilometers, 38 dilapidations, covering 75.6 square kilometers, and 9 mud-rock flows covering an area of 23.9 square kilometers.

The overlap rate between two adjacent high-resolution images of the disaster area from airborne remote sensing is high. Using this characteristic and remote sensing image processing, the images can generate red and blue stereopairs. Through red and blue stereo glasses, they can show the 3D Spatial Information of buildings in the disaster area intuitively and greatly improve the identification and extraction efficiency of housing damage information. The identification accuracy of semi-damaged buildings increases to more than 10%.

(a)   (b)

Fig. 8. (a) 3D analysis of secondary geological disasters in the Wenchuan earthquake, (b) 3D view in red-blue mode for damaged building extraction.

5. Conclusion

Remote sensing monitoring of the Wenchuan and Yushu earthquakes' secondary geological disaster shows that high-resolution optical remote sensing, which can extract the seismic secondary disaster remote sensing characteristics accurately and monitor and evaluate the information of spatial distribution, damage degree and so on of earthquake secondary geological disasters has some advantages, such as intuition, large information and quantification. SAR has the advantages of all-weather data acquisition. The Yushu earthquake multi-mode SAR remote sensing monitoring study has proven that multi-mode SAR is effect and has important potential in earthquake disaster analysis and evaluation. Three-dimensional computing technology for measuring secondary geological disasters is an important technology, which not only improves the calculation and simulation accuracy of secondary geological disasters, but also can promote collaboration on three-dimensional simulation technology and auxiliary mitigation and provide analysis platforms for interactive operation in secondary geological disasters.

In addition, quantitative and reliable evaluation of secondary earthquake disaster depends on high-resolution Earth observation technology. But at present the automatic disaster monitoring algorithms and software for high-resolution Earth observation images still cannot meet actual needs, and 3D interactive analysis platform technology is still not mature. Meanwhile, secondary geological disaster monitoring relies heavily on traditional man-machine interactive visual interpretation technology.

6. Acknowledgement

The authors would like to thank all of the team members who participated in the Wenchuan Earthquake Disaster Reconstruction, Monitoring and Assessment Using Remote Sensing Technology.

This work was supported by the National Basic Research Program of China (2009CB723906, 2009CB723902) and National Natural Science Foundation of China (60972141).

7. References

Ainsworth T L, Schuler D L, Lee J S. Polarimetric SAR characterization of man-made structures in urban areas using normalized circular-pol correlation coefficients. Rem Sens Environ, 2008, 112: 2876–2885

Chen Yuntai, Xu Lisheng, Zhang Yong et al., Report on the Wenchuan large earthquake source of May 12, 208, http://www.csi.ac.cn/Sichuan

Cloude S R, Pottier E. A review of target decomposition theorems in radar polarimetry. IEEE Trans Geosci Rem Sens, 1996, 34: 498–518

Cloude S R, Pottier E. An entropy based classification scheme for land applications of polarimetric SAR. IEEE Trans Geosci Rem Sens, 1997, 35: 68–78

Cui Peng, Wei Fangqiang, He Siming et al. 2008. Mountain disasters induced by the earthquake of May 12 in Wenchuan and the disasters mitigation. JOURNAL OF MOUNTAIN SCIENCE, 26(3):.280-282(In Chinese)

Deng Qidong, Ran Yongkang, Yang Xiaoping. 2007. Active tectonic map of China. Beijing: Earthquake Press (In Chinese)

Ermini L. and Casagli N. 2003. Prediction of the behavior of landslide dams using a geomorphological dimensionless index. Earth Surface Processes and Landforms, 28(1): 31-47.

F.Mattia, T.Le Toan, J. C.Souyris, G.D.Carolis, N.Floury,and F.Posa, The effect of surface roughness on multifrequency polarimetric SAR data . *IEEE Trans Geosci Rem. Sens.*, vol. 35, no. 4, pp. 954–966, July 1997

Fu Bihong, Shi Pilong, Zhang Zhilong. 2008. Spatial characteristics of the surface rupture rroduced by the MS 8.0 Wenchuan earthquake using high-resolution remote sensing imagery. ACTA GEOLOGICA SINICA, 82(12): 1679-1687(In Chinese)

Ge Yong, Xu Jun, Liu Qingsheng, et al. 2009. Image interpretation and statistical analysis of vegetation damage caused by the Wenchuan earthquake and related secondary disasters. Journal of Applied Remote Sensing, 3, 031660.

Guo H D, Liao J J,Wang C L et al. 1997. Use of multifrequency,multipolarization shuttle imaging radar for volcano mapping in the kunlun mountains of western China. Remote Sensing of Environment, 59:364-374.

Guo H D, Zhu L P,Shao Y et al. 1996. Detection of structural and lithological features underneath a vegetation canopy using SIR-C/X-SAR data in Zhao Qing test site of southern China. Journal of Geophysical Research, 101(E10): 23101-23108.

Guo Huadong, Liu Hao, Wang Xinyuan, et al. 2000. Subsurface old drainage detection and paleoenvironment analysis using spaceborne radar images in Alxa Plateau, Science in China Series D: Earth Sciences, 43(4): 439-448.

Guo Huadong, Liu Liangyun, Lei Liping, et al. 2010a. Dynamic analysis of the Wenchuan Earthquake disaster and reconstruction with 3-year remote sensing data, International Journal of Digital Earth, 3(4):355–364.

Guo Huadong, Wang Xinyuan, Li Xinwu, et al.. 2010b. Yushu earthquake synergic analysis using multimodal SAR datasets. Chinese Science Bullet, 55(31): 3499-3503.

Guo Huadong. 2000. Radar for Earth observation: Theory and Applications, Beijing: Science Press(In Chinese)

Han Yongshun, Liu Hongjiang, Cui Peng, et al. 2009, Hazard assessment on secondary mountain-hazards triggered by the Wenchuan earthquake. Journal of Applied Remote Sensing, 3, 031645.

Huang Xiaoxia, Wei Chenjie, and Li Hongga. 2009. Remote sensing analysis of the distribution and genetic mechanisms of transportation network damage caused by the Wenchuan earthquake. Journal of Applied Remote Sensing, 3, 031650.

Jin Feng, Shen Xuhui, Hong Shunying, Ouyang Xinyan. 2008. The application of remote sensing in the earthquake science research. Remote Sensing for Land&Resources, (2): 5–8(In Chinese)

Lei Liping, Liu Liangyun, Zhang Li, et al. 2009. Assessment of spatial variation of the collapsed houses in Wenchuan earthquake with aerial images. Journal of Applied Remote Sensing, 3, 031670.

Li ping, Tao xiaxin, Yan shiju. 2007. 3S-based quick evaluation of earthquake damage, Journal of natural Disasters, 16(3):110-113 (In Chinese)

Liou Yuei-An, Kar Sanjib K.; Chang Liyu. 2010. Use of high-resolution FORMOSAT-2 satellite images for post-earthquake disaster assessment: a study following the 12 May 2008 Wenchuan Earthquake. International Journal of Remote Sensing, 31(13):3355 – 3368.

Liu Liangyun, Wu Yanhong, Zuo Zhengli, et al.. 2009. Monitoring and assessment of barrier lakes formed after the Wenchuan earthquake based on multitemporal remote sensing data. Journal of Applied Remote Sensing. 3, 031665.

Massonnet D., Feigl K. 1998. Radar interferometry and its application to changes in the Earth's surface. Reviews of Geophysics, 36(4):441-500.

Paul Rosen, Scott Hensley, Ian R. Joughin, et al. 2000. Synthetic aperture radar interferometry. Proceedings of the IEEE, 88(3):333–382

Ramesh P. Singh. 2010. Satellite observations of the Wenchuan Earthquake, 12 May 2008. International Journal of Remote Sensing, 31(13):3335 – 3339.

Wang Erqi, Zhou Yong, et al. 2001. Geologic and geomorphic origins of the east Himalayan gap. Chinese Journal of Geology (SCIENTIA GEOLOGICA SINICA), 36(1):122-128(In Chinese)

Wang Shixin, Zhou Yi, Wei Chengjie, ShaoYun, Yan Fuli. 2008. Risk Evaluation on the secondary disasters of dammed lakes using remote sensing datasets , in the'Wenchuan Earthquake. Journal of Remote Sensing, 12(6):900-907(In Chinese)

Xiao lebin, Zhong ershun, Liu jiyuan, Song guanfu, 2001, A Disussion on Basic Problems of 3D GIS, Journal of Image and Graphics, 6(9): 842-848. (In Chinese)

Xu Min, Cao Chunxiang, Zhang Hao, et al. 2010. Change detection of an earthquake-induced barrier lake based on remote sensing image classification. International Journal of Remote Sensing, 31(13): 3521 – 3534.

Xu Weihua, Dong Rencai, Wang Xuezhi, et al. 2009. Impact of China's May 12 earthquake on Giant Panda habiat in Wenchuan County. Journal of Applied Remote Sensing, 3, 031655.

Zhang Wenjiang, Lin Jiayuan, PengJian, et al. 2010. Estimating Wenchuan Earthquake induced landslides based on remote sensing. International Journal of Remote Sensing, 31(13): 3495 – 3508.

Zhang Yong, Xu Lisheng, Chen Yuntai, et al. 2010. Fast inversion of rupture process for 14 April 2010 Yushu, Qinghai, earthquake. Acta Seismologica Sinica, 32(3): 361-365(In Chinese)

Zhuang JianQi, Cui Peng, Ge YongGang, et al. 2010. Probability assessment of river blocking by debris flow associated with the Wenchuan Earthquake. International Journal of Remote Sensing, 31(13):3465 – 3478.

4

Newly-Proposed Methods for Early Detection of Incoming Earthquakes, Tsunamis & Tidal Motion

Samvel G. Gevorgyan

*Center on Superconductivity & Scientific Instrumentation, Chair of
Solid State Physics, Faculty of Physics, Yerevan State University;
Institute for Physical Research, National Academy of Sciences;
Precision Sensors/Instrumentation (PSI) Ltd.
Armenia*

1. Introduction

During a last decade scientists and engineers step-by-step are developing a Single-layer Flat-Coil-Oscillator (**SFCO**)-based *new measurement technology,* and looking for its effective use in a research, and elsewhere. It was introduced in 1997 by our group in Armenia [1-2] and then improved by an integrated research group in Kyushu University, Japan, during next 4 years (1998-2002) [3-4] – allowing to reveal fine physical effects related with the basic properties of high-T_c superconductors (**HTS**) [5-8]. Starting with 2004 the method passed further development in Armenia, and was then applied for creation of a new *absolute*-position sensor of *nano*-scale resolution [9]. Advantages of the *SFCO* method-based position sensor become more evident when applied to the *quasi*-static Seismometry – to study slow movements of ground. Due to these, the *SFCO measurement technology* (in a whole [1-4]), and its first application as a novel seismic detector of slow movements (in particular [9-10]) appeared among the Top six World Security Technologies at the 2008 year's *"Global Security Challenge"* competition – details on *"GSC-2008"* forum see in: http://www.globalsecuritychallenge.com. In this Chapter, we discuss principle of operation, and test data of such a new *absolute*-position sensor, installed (for validation) in a well-known seismometer, as an additional pick-up component – showing its advantages compared to traditional technique. We discuss also wide potential of this new method, as a real-time measurement technique for early detection of incoming earthquakes, tsunamis and tidal motion. We also outline prosperous future of such a sensor. To sense what are advantages of the flat-coil-based this unique method, let's remember: oscillators are among the most of precise measuring instruments, because the frequency is possible to measure with a very high accuracy. Among them, those at *MHz* frequencies, having volume pick-up coils (mainly, solenoid-shaped), activated by a *low*-power (*backward*) tunnel diodes (**TD**) (see [2, 11-12] and references therein), are of special interest. Replacement of such a standard coil by the unusual, single-layer flat (open-faced) one, as a detecting circuit in a stable-frequency and amplitude *TD*-oscillator, enabled to make coil's filling factor close to the maximal possible value (the *unit*) for flat objects, resulting in strong enhancement of the resolution of measurements by 3–4 orders of magnitude (especially, in studies of thin, plate-like *HTS* materials [1, 3-8]). For comparison, typical values of the filling factor for solenoid coils are 10^{-4}–10^{-3} for the said samples. Advantages of the *SFCO* technique become more evident at slow movements of the objects, positioned near the coil face. Just therefore, this method has been very soon applied for the creation of a *nano*-scale *absolute*-shift position sensor, which one may successfully use in many areas: for example, for the *quasi*-static (*slow-movement*) Seismometry [9-10], in various security systems. Why this problem is so urgent? Basically, there are two types of seismic sensors, acting presently [13]: *inertial seismometers*, which measure ground motion relative to some inertial reference (*suspended inert mass*), and *strain-meters* (or *extensometers*), which detect shift between two points of the ground. Although strain-meters are conceptually simpler than inertial seismometers, their technical realization is much more difficult. Besides, as ground motion relative to the suspended inert mass is usually larger than differential motion within a test tube of reasonable dimensions, inertial seismometers usually are more sensitive to earthquakes. At low (and especially, at *super*-low) frequencies, however, it becomes hard to maintain the hanging reference fixed, and for detection of *quasi*-static deformations and *low*-order free oscillations of the earth's crust, tidal motion (*moon movement*), and for observation of mechanical vibrations of buildings, bridges, etc., the strain-meters may take noticeable lead over inertial seismometers. We describe in this Chapter how to overcome such lack of acting seismographs/accelerometers/vibrometers by the use of the recently offered by us flat-coil-based, super-broadband, nano-scale-resolution position sensor [9-10]. The more so, because further development of such a highly sensitive sensor technology may contribute also to on-time tracking (*prediction*) of potential incoming tsunamis, and monitoring of the state and zone borders as well.

2. Flat coil-based *absolute*-position sensor for *nano*-scale resolution, *super*-broadband Seismometry

And so, a new class *super*-broadband, *nano*-scale resolution position sensor is developed and tested by our group. It can be used, in particular, as an additional sensor in presently acting seismographs. It enables to extend *frequency*-band (theoretically, up to "zero"), and enhance *absolute*-resolution (*sensitivity*) of seismographs available on the market (*by at least an order of magnitude*). It allows transferring of the mechanical vibrations of constructions, buildings, bridges & ground with amplitudes over *1nm* into detectable signal in a *frequency*-range starting practically from the *quasi*-static movements ("zero"!). It is based on detection of position changes of a vibrating normal-metallic plate placed near the coil face – being used as a pick-up circuit in a stable *TD*-oscillator. Frequency of the oscillator is used as a detecting parameter, and the measuring effect is determined by a distortion of the *MHz*-range testing field configuration near the coil face by a vibrating plate, leading to magnetic inductance changes of the coil, with a resolution *1-10pH* (*depending on operation temperature of a technique*). This results in changes of test oscillator frequency. Below, we discuss work-principle, and test data of such a new position sensor, installed in a known Russian *SM-3* seismometer (for validation) as an additional pick-up element – showing its advantages compared to traditional techniques. We also discuss potentials of this novel *absolute*-position sensor, operating down to liquid-^{4}He temperatures, and in high magnetic fields – as a real-time measurement element for early detection of earthquakes, incoming tsunamis, tidal motion, and for tracking borders. We discuss also possible design of seismic detectors based on this sensor. Besides, we outline perspective future of such an unprecedented sensor – involving substitution of a normal-conducting pick-up coil by a superconductive one, and replacement of a tunnel diode by the *S/I/S hetero*-structure – as much less-powered active element in a detecting oscillator, compared to the tunnel diode. These may improve stability of oscillators, created by the use of *SFCO* method, and thus, enhance the resolution of seismic devices, and tsunami detectors as well – by at least another 2-3 orders of magnitude. Such improvements may enable to reveal and study *quasi*-static deformations and *low*-order free oscillations of earth's crust, precursor to earthquakes. It may also permit to study features of the tidal motion and tsunami waves. Such a sensor may be also used as a *position/vibration* sensing element in *micro*- and *nano*-electronics (in probe microscopy), in security systems, and in medicine as well.

2.1 Traditional inertial seismometer

A *Traditional inertial seismometer* converts ground motion into electrical signal, but its properties cannot be described by a single-scale parameter, such as the output volts per millimetre of the ground motion [13] (*as occur in case of the absolute*-position *sensors*). Its response to ground motion depends not only on the amplitude of motion (*how large it is*) but also on its time-scale (*how fast it is*). So, the suspended (*hanging*) seismic mass has to be kept in place by certain restoring force (*electromagnetic, mechanical, else nature*). But, when ground motion is slow, the mass will move with the body of a seismometer, and the output signal even for a large motion will thus be negligibly smaller. Such a system is so a high-pass filter for ground shifts. This must be taken into account if the ground motion is reconstructed from the recorded signal. So, creation of seismic detectors, which may give large output both for fast and slow motion (*regardless of the rate of motion – as absolute-position sensors behave themselves*), still remains among the prime important problems in the Seismology (*and not only…*).

2.2 Principle of operation of new seismic detector

To this end, a prototype of the *SFCO* method-based position sensor has been created and installed by us in a setup of the Russian seismometer of *SM-3* type (Fig.1a). In such a *"hybrid SM-3"* device (Fig.1b) a flat coil serves as a pick-up in a stable *16MHz*–oscillator, driven by a *low*-power Russian tunnel diode of the *AI-402B* model. Actually, 2 similar flat-coil oscillators are mounted in *SM-3*. One is used as a position detector, the other – to detect background at all times (*bottom* and *top* oscillators in Fig.1b respectively). Let-in *SM-3* position sensor is extra to its own *vibro*-sensor one, based on excitation of the electro-motive force (**EMF**) in a solenoid coil (Figs. 1a and 1c). In case of the *SFCO*-based sensor, measuring effect is proportional to changes of mutual distance between the coil and metallic plate vibrating parallel to the coil face (d in Fig.1c). This results in the changes of the *test*-oscillator frequency.

So, new seismic detector converts ground motion into shift of a flat-coil-oscillator frequency – *due to ground shaking*. The measuring signal appears as a result of the coil motion (fixed on seismograph's body – Figs. 1b-1c and 2) relative to metallic plate (fixed on hanging pendulum (Fig.1c), or membrane (Fig.2)), positioned near the coil. Figures. 1c and 2 schematically illustrate *SFCO* sensor-based novel seismic detectors' possible designs: F_S is the shock force, and d – amplitude of vibration of a pendulum (see Fig.1c) or membrane (Fig.2), caused by it.

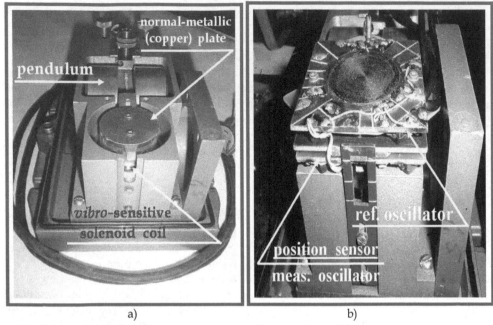

a)   b)

Fig. 1. a) Top view of the *original Russian SM-3 seismograph*, with the light metallic (*copper*) plate additionally screwed on its vibrating pendulum (*schematics see in* Fig.1c). Initially, the *SM-3* device is designed to detect vibrations in a frequency range from *0.5Hz*, and up to *50Hz*.
b) Front-view of the *original Russian SM-3 seismograph*, with additionally installed package with 2 flat-coil-based oscillators – named as the *"hybrid SM-3"* seismograph.

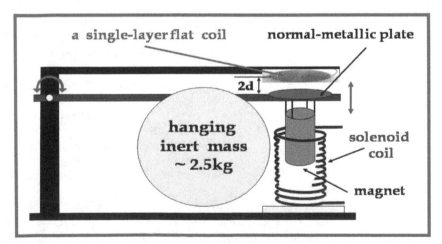

Fig. 1. c) Mechanical schematics of the *"hybrid SM-3"* seismograph – advanced by the use of *SFCO* method-based highly sensitive, *super*-broadband position sensor: *d* is the amplitude of vibration of a pendulum, caused by the ground shaking.

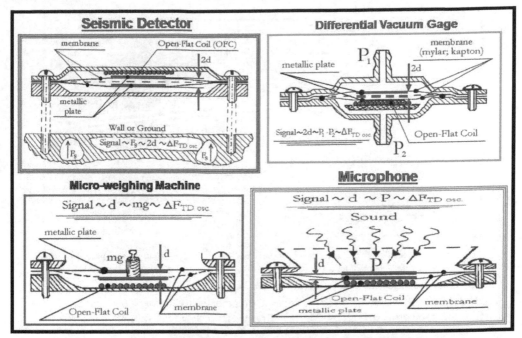

Fig. 2. Mechanical schematics of the *SFCO* sensor-based fully novel 4 techniques: seismic detector, differential vacuum gage, microphone, as well as micro-weighing machine: F_S is the shock force, d – the amplitude of flapping of a membrane, caused by the ground shaking.

2.3 Flat-coil based measurement technology: Its advantages

A single-layer flat-coil-oscillator test method (the **SFCO** technique [1-2] – it is introduced by our group in 1997, its electrical scheme is shown in Fig.3) is a fine research instrument for doing *MHz*-range, sensitive measurements. It can be used for determination of too much little changes of distances with $\Delta d \sim 1\text{-}10\text{Å}$ absolute and $\Delta d/d \sim 10^{-5}\text{-}10^{-6}$ relative resolution (depending on a model and working temperature of the *TD*-oscillator [3-4]. It is also a sensitive radio-frequency (**RF**) Q-meter – to study absorption as small as $10^{-9}W$ in thin flat materials (for example, in plate-like high-T_c superconductors [5-7]). The *SFCO* method can operate down to the liquid-^{4}He temperatures. Presently, it is tested by us up to $12T$ magnetic fields [7]. The method differs from the known "*LC*-resonator" technique (see, for example, [14]) by replacement of the volume-shaped testing coil by the unusual single-layer flat (open-faced) one. Additionally, it is driven by the *stable*-frequency, *low*-power tunnel diode.

Advantages of the *SFCO* method-based sensor become more evident when applied to *quasi*-static Seismometry – to study slow movements of ground. In this regards, Fig.4 compares responses of the *SFCO* position sensor and the *EMF*-based *world*-best *SM-24 ST vibro*-sensor (geophone.com) – against the same vibrations. The vertical size of the blacked-out region in this Figure shows advantages of our novel *SFCO*-sensor for different values of vibration frequencies. One may conclude from the Fig.4, that advantages of the *SFCO* method-based new sensor become much more evident at *super*-slow vibrations (*movements*), with F< 10Hz.

Both the frequency and amplitude of the oscillator are used as testing parameters in a *SFCO* technique. The measuring effects are determined by a distortion of the coil testing field

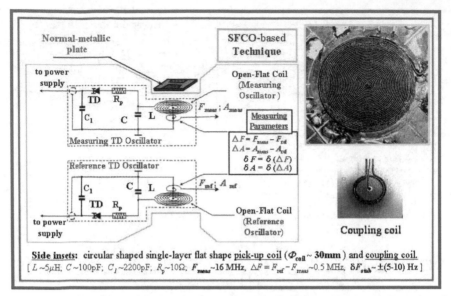

Fig. 3. Electrical schematics of the new seismic detector, based on *SFCO* technique (single-layer flat-coil-oscillator, driven by the *stable*-frequency, *low*-power tunnel diode (**TD**).

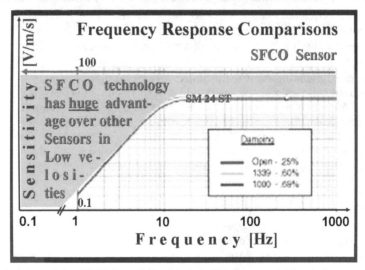

Fig. 4. Comparision of the *SFCO*-based *absolute*-position sensor with an electro-motive-force (**EMF**)-based *world*-best *SM-24 ST vibro*-sensor (geophone, see: www.geophone.com).

configuration near its flat face, and by the absorption of the same field's power by an object under test (*due to external influences*). These finally result in the changes of test oscillator frequency and/or amplitude, respectively. Compared with the traditional (*volume*-coil) method, in a flat-coil technique testing *RF*-field is densely distributed near the coil face. Besides, due to flat shape, even a little shift of the position of a *normal*-conducting plate, placed near the coil, may lead to strong distortion of the field distribution around the coil. These features, and the stability of *TD*-oscillators ($\Delta F_{stability} \sim 1\text{--}10Hz$, $\Delta F/F \sim 10^{-7}\text{--}10^{-6}$ – depending on the model & temperature – see [2, 11-12]) enabled us to reach 6 orders relative resolution in *SFCO* technique [3-4], permitting to effectively use it in a basic research [5-8], as well as in some modern technical applications [9-10]. In the last case, frequency of the oscillator is mainly used as a testing parameter, and the measuring effect is determined by a distortion of the *MHz*-range testing field configuration near the coil face by the vibrating copper plate, leading to the magnetic inductance changes of the coil, with a resolution ~ *1-10pH* (*depending on operation temperature of a technique*), resulting in the changes of test oscillator frequency.

2.4 Reconstruction of ground motion from recorded frequency-shift of *TD*-oscillator

Since electro-motive force based traditional *vibro*-sensors (included, the own sensor of *SM-3*) and suggested by us position sensors are various nature devices, with different outputs (*EMF*-based sensor converts ground motion into output volts, while flat-coil-based novel sensor converts the same motion into the shift of test-oscillator frequency), there are no direct ways to compare them properly, except that one may compare their responses over the respective noises during the same shaking. And so, we tried to detect and compare signal-to-noise (**S/N**) ratios for these two (*different principle of operation*) sensors, during the same experiment – against the same *1-2Hz* time-scale weak vibration.

In this regard, note that for correct reconstruction of the ground motion from the recorded frequency shift there is need to properly calibrate the *SFCO* method-based this non-traditional technique. The problem here is much complicated compared with the cylindrical (*solenoid*)-coil based technique, since even for the simplest case of a weakly vibrating thin conducting plate near the flat coil the calibration data are dependent on the used plate's diameter. For comparison, in case of cylindrical (*solenoid*) coil-based similar technique one needs calibration for only one (*given volume*) cylindrical sample, placed in the homogeneous testing field area inside the coil. Then, the obtained *calibration*-data can be expanded and used for any other shape and volume samples, provided that they are positioned anywhere inside the almost homogeneous-field area, near the cylindrical coil center [14].

So, below we discuss briefly the method, and results of calibration of the tested flat coil's *RF*-field configuration, by the use of a normal-conducting (*copper*) plate enabling correct transfer of the measured shifts of frequency δF, to the changes of distance δd, from the coil face *d*. One of possible ways to do that seems the calibration by moving the *given*-size *disk*-shaped copper plate towards the coil's face, up to the given distance, *d*, and back. This strongly changes the coil's testing field configuration (*and thereby, oscillator frequency*), and enables the empirical estimation of the so-called *G*-factor – as the coefficient for the coil's inductance (*resonant frequency*) modulation. Changing the position of the metallic object, we could experimentally determine the value of the *G*-factor as the relation between the resonant frequency modulation δF and the change in position δd. Figure 5 presents and illustrates the results of such calibration of the created position sensor (*let-in the SM-3 seismic device*) – which we realized. As is seen, the empirically determined *G(d)*-factor (*which actually is the absolute resolution of the technique*) for the given area metallic plate depends on the position *d*, near the flat coil. *G*-factor enables correct transferring of the measured shifts in frequency to the linear changes in distance by the formula: $\delta d \equiv -G(d) \times \delta F$, important for the proper reconstruction of the ground motion from the recorded *frequency*-shifts. Figure 5 shows that *G*-factor depends strongly on distance from the coil face. Namely, sensitivity (*absolute resolution*) drops exponentially with an increasing distance – due to sharp drop of a testing field density. $G_w \sim 1\mathring{A}/Hz$ in Fig.5 is a typical value of a geometric factor achieved for the $F_{meas} \sim 16MHz$ operating frequency and $\Phi_{coil} \sim 30mm$ coil oscillator on $d \sim 1.1mm$ distance from the coil face, at liquid–^{4}He temperatures (*typical stabilities reached for TD-oscillators at low temperatures are* $\delta F_{stability} \sim \pm$ **(1-2)** *Hz* – see Fig.6b) [3-4, 15]. At the room temperatures, the noise level of the tested flat-coil sensor (*let-in the SM-3 seismic device*) is a little bit worse – close to ± **(5-10)** *Hz*.

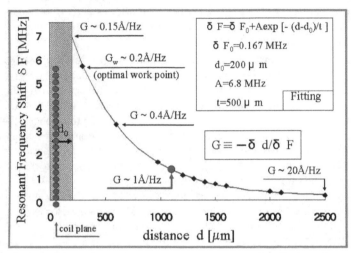

Fig. 5. *SFCO* technology-based position sensor sensitivity vs. the distance from the coil flat face: testing *RF*-field's density vs. the distance from the open-flat coil's face in a *SFCO* position sensor.

Note that such a low noise level of the tested sensing system is due to changes in inductance caused by all internal factors in the system's electronics, and mechanics. To be sure in this matter fully, we fixed mechanically (*for a long time*) the pendulum of the *"hybrid SM-3"* (see Figs. 1a - 1c), and tried to detect noise level of the measuring oscillator. Its stability was close to ± **(5-10)** *Hz* at room temperatures, during an hour. And so, distance d can be taken as a unique factor to determine inductance changes in measurements (*due to vibration of a copper plate near the coil face*) in the range of resolution corresponding to the frequency shift of about ± **(5-10)** *Hz*, at room temperatures. Hence, for the $\Phi_{coil} \sim 30mm$ coil sensor, installed in *"hybrid SM-3"* (*with the copper plate, vibrating near the coil face, at a distance* $d \sim 1.1mm$), we reached a re-solution $\delta d = G \times \delta F_{stab} \sim 1\mathring{A}/Hz \times \pm$ **(5-10)** *Hz* $\sim \pm 1nm$ at the room temperatures (see Fig.5).

2.5 Novel seismic detector based on SFCO measurement technology (test-results, discussion, future perspectives)
2.5.1 Test-results
Thus, because there is no other reasonable ways for direct comparison of the said 2 different nature (*principle of work*) sensors we tried to compare their responses over respective noises, during the same shaking. So, we detected, and below compare, the signal-to-noise ratios for above sensors – during the same experiment, against the same *1-2Hz* time-scale weak vib-ration. Comparative-test data of such an experiment are shown in Fig.6. In our tests, the *"hy-brid SM-3"* was fixed to the glazed-tile floor of a laboratory room, situated on the 2-nd floor.

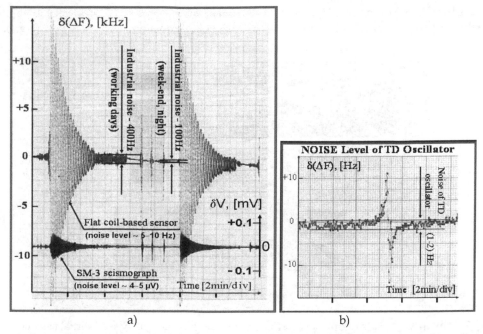

Fig. 6. a) Comparative-test data of the flat-coil-oscillator based *absolute*-position sensor (left vertical scale – δ(ΔF), [kHz]) and *EMF*-based *vibro*-sensor (right vertical scale – δV, [mV]) – both installed in the same *"hybrid SM-3"* seismic device. Room-temperature noise levels of both sensors are also pointed out in the figure (~ 5-10Hz and ~ 4-5μV, respectively).
b) Noise level (*stability*) of a tested *TD*-oscillator at liquid–⁴He temperatures, permitting to estimate an extreme resolution one may reach in *"hybrid SM-3"* seismic device, supposing that its *SFCO* novel position sensor is cooled down to 4K. Note, that the room-temperature noise level of the tested *SFCO* sensor is a little larger – close to ±(5-10)Hz. The room-temperature noise of the *SM-3's EMF*-based own *vibro*-sensor is about ~ 4-5μV – see Fig.6a.

First, from data shown in Fig.6a one may conclude that, as detected by a *SFCO* position sensor, the level of background vibrations of a laboratory floor is near ±400Hz – during workdays. Taking into account the above said value of about 1Å/Hz for the G-factor at d~1.1mm *work*-distance from the coil (see Fig.5) such level of background vibrations corresponds to the amplitude of vibration of the laboratory floor of about ±40nm. Besides, Fig.6a indicates that background vibrations of our laboratory building were almost 4 times stronger at workdays, compared to weekends and nights. Even such shakings at nights, however, almost *50* times exceeds the measured noise level (of about ±1-2Hz – Fig.6b) one may get in created *"hybrid SM-3"* seismic device – provided that its *SFCO* position sensor is cooled down to 4K. Background shakings of the laboratory room might be caused by the industrial pumping of an environment, and besides, by the vibration of earth's crust. Background shakings might be caused also by rocking during the tests of a technical nature. In this regard, note that a fine signal, seen in Fig.6b, detected by our *SFCO* method-based new sensor, is an evidence of its high abilities. The signal is result of beating of the measuring *TD*-oscillator with a little signal "coming" from the close-located broadcasting station. An acting seismic station is under creation in Yerevan State University, based on created *"hybrid SM-3"* new seismographs, capable of providing LabVIEW environment-based data acquisition and processing (Fig.7).

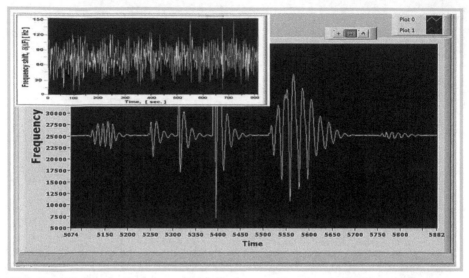

Fig. 7. LabVIEW signals of our new *SFCO absolute*-position sensor-based inertial seismic detector (the *"hybrid SM-3"* seismograph) for different amplitude shakings, ranging from ±25 to ±250nm, at the background vibration of about ±5nm (see inset on *top left*). Background-vibration LabVIEW signals of the *SFCO-* sensor based new inertial seismic detector. Experiments were conducted at the night time-period, to achieve as low as possible noise level at room temperature in a technique caused by the industrial rocking of an environment and vibration of the earth's crust.

2.5.2 Discussion

Comparison of signal-to-noise ratios (at F~1Hz), for new sensor (*flat-coil based SFCO sensor* – $(s/n)_{flat-coil}$ is about 16kHz/(5-10Hz) ≅ 1600-3200) and for *SM-3* sensor ($(s/n)_{EMF-sensor}$~150μV /(4-5μV) ≅ 30-35) – both operating in the same *"hybrid SM-3"* seismograph – permits to conclude that the *SFCO* sensor is more sensitive by about 50-100 times (see Figs. 6a and 8). Besides, since the *SFCO* sensor allows detecting of *absolute*-position shifts (see Fig.4, low frequencies), it may enable to detect very beginnings of *quasi*-static deformations and oscillating processes in earth crust – at very low frequencies – in contrast to the traditional *EMF*-based sensors, being used practically in all acting inertial seismometers of a different design. This is the case since *EMF*-sensor may not detect slowly passing processes – due to minor voltage arising in solenoid pick-up coils during the slow movements of a pendulum (Fig.1c). So, in order to effectively detect *quasi*-static deformations by the *SFCO* technology-based *absolute*-position sensor, one should build and use a properly vibrating mechanical pendulum (*with a mass as heavy as possible*, and *with as weak as possible restoring force of the mechanical part of pendulum*) – something like to what is the case in Russian *SM-3* detector, but with less friction against the motion of a freely hanging pendulum. *EMF*-based sensor may not detect slow processes, at any case, since it is a velocity sensor. This all may become crucial for detection of low-order free oscillations of the earth crust, and for observation of the peculiarities of a few-hour duration tidal motion & tsunami shaping. That is why one should use the *SFCO absolute*-position sensing technology (in this, or another modification of a sensor – *see schematics of different sensors in Fig.2, to be used depending on the application*) to reveal in advance, and study origins of formation of earthquakes, tsunami waves, and tidal motion – *impossible, in principle, for other methods*. We believe this offer holds considerable potential for meeting advanced technical needs of the seismic & tsunami services supported by governments of practically all countries positioned in the seismically active regions of the world.

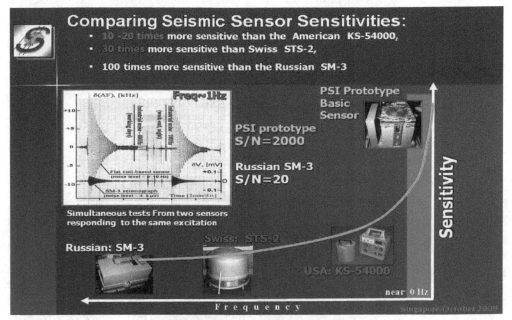

Fig. 8. Comparision of *"hybrid SM-3"* seismic detector (based on a *SFCO* technology *absolute* position sensor), with the *EMF* principle of operation based other word-wide detectors.

In this connection, we bring in a next Fig.8 comparative data, related with the *SFCO absolute*-position sensor technology-based *"hybrid SM-3"* and the *EMF*-based word-wide seismic detectors. Comparison is again made at vibration with F~1Hz. Taking into account huge advantages of the *SFCO* position sensor technology over the other sensor technologies (especially, at vibrations with F < *10Hz* – see Fig.4) much higher sensitivity of the said *"hybrid SM-3"* detector (having inside integrated *SFCO* sensor, as the additional sensing element) becomes evident. As to the vibration frequencies below the 1Hz, the *EMF*-based all seismic sensors loss their row sensitivity at al (sensitivity, without long-time and expensive integrating electronics) – see Fig8 and Fig.4.

2.5.3 Future perspectives
There are many ways how to even more enhance the resolution of such new *absolute*-position sensors, and, as a result, capabilities of the presently acting *inertial seismometers* – even by the several orders of magnitude. For that purpose, the pick-up flat coil, and/or the active element of the measuring oscillator should be made of superconductive material (high-T_c or low-T_c – for better stability). In other words, one of the relatively easier ways relates with

the replacement of the normal-metallic coil by the superconductive one. This may improve the tunnel diode oscillator stability by at least 1-2 orders of magnitude [2]. The next improvement relates with the substitution of the tunnel diode by the superconductive *S/I/S* hetero-structure – as much more less-powered active element (*compared to tunnel diodes*) for the measuring oscillator of the *SFCO absolute*-position sensor, with a few orders of magnitude less steep of its *I-V* curve's negative differential resistance [16]. This may raise the oscillator stability by another 2-3 orders of a value [2]. Even these two modernizations are enough in order to enhance the stability of the measuring and reference oscillators of such a technique (Fig.3) and hence, to increase the signal-to-noise ratio (*sensitivity*) of the *SFCO* technology-based seismic detectors – by at least 3-4 orders of a value. As follows from the Fig.5, the *absolute*-resolution of such a new sensor drops exponentially when a normal-conducting plate moves away from the coil face. This property of *SFCO* sensors makes easy adjustment of the sensitivity (*resolution*) of such a new position sensor, for various practical usages in future.

3. Areas for specific application of SFCO *absolute*-position sensors

Besides the usage of the SFCO technology - based *absolute*-position sensors in seismic predicttion & protection, they might be also effectively applied in: *security systems; geophysics & town-planning; micro- & nano-electronics; military science, engineering & intelligence;* etc.:

Fig. 9-10. The *SFCO*-sensor based *Early Warning Security System* can secure the runway and specific underwater perimeter with the invisible and totally passive security net, and can detect over the ground and underground, as well as underwater moving intruders.

in security systems: The new (**SFCO**) technology *absolute*-position sensor-based ultra-sensitive seismic detectors and vacuum gages may give rise to many markets & applications, and bring to products that can serve both military & civilian applications. Early warning security systems (**EWSS**) are natural applications that can serve to protect State & Federal borders, provide Ports security & control, as well as Civilian applications of perimeter security

Fig. 11. The *SFCO*-sensor based *Early Warning Security System* can secure the ground and underground, as well as specific underwater perimeter with the invisible and totally passive security net, and identify the location of underwater moving intruders.

controls such as security of oil pipelines, airports, and private properties. The new technology sensors may also enable detection and recognition of various mobile targets (*walking or crawling man, vehicles, tanks, or other human activities*) approaching any zone (*military camps, state properties, banks, or other critical high priority infrastructures*) or borders without the need of physical line of sight.

Figures 9 through 11 are pictorial depictions of systemic applications to real world security scenarios, showing the flexibility and versatility of this new technology rendering one of the highest quality EWSS for military or civilian applications, covering detection for underground movements, over the ground movements, and underwater movement.

in geophysics and town-planning: for gas and oil prospecting, and also to reveal too much weak vibrations and slow bending (*twist*) of the buildings, constructions and bridges, as well as for permanent monitoring of old bridges aging;

in micro- and nano-electronics: for creation of *New Generation* microscopes with long-range action "*magnetic-field*" probes.

Our recent research shows [17-18], that flat coil based *TD*-oscillators can be activated also with their internal capacitances (*without an external capacitance C in their resonant circuits* – see **Fig.3**). That is the result of relatively high value of internal capacitances of single-layer flat coils compared to their parasitic capacitances with respect to the surrounding radiotechnical environment. This opens one more exotic area for flat-coil oscillator application. Namely, a "*needle-like*" testing magnetic field of such a flat coil (see **Fig.12a**), used as a pick-up in such a stable *TD*-oscillator, enables a novel method (*new approach*) for surface probing, based on re-placement of short-range, solid-state probes of acting microscopes (*such as needles or cantilevers of tunneling [19-20] and atomic-force [21] microscopes, probes of the near-field microscopes,* etc.) by the long-range action non-solid-state ones. Such an unusual probe shows strong dependence of a detected signal on the size of the spatial-gap between the probe and the surface of the object – crucial for the probe microscopy (**PM**) [22]. This opens an opportunity for creating of the "*magnetic-field*" probes with a *RF* power applied to the sample lying in the range of 1nW to 5μW. The gap between such a "*probe-formative*" flat coil and the object can be larger than 100μm [18], compared with the 1nm gap of the acting probe microscopes [22]. In our tests we reached a lateral resolution ~ 1μm even for the relatively large diameter ($2R_{coil}$ ~ 14mm) flat-coil technique [18].

Such a *SFCO*-probe may also "*notice*" and distinguish details of the relief of the normal-metallic object – with about 10μm spatial-resolution, presently (**Fig.12**). In order to demonstrate that, we performed an experiment with one-dimensional (**1D**) metallic grid made of 6 copper wires (see **Fig.12a**): each wire was ~20-30μm in dia. and was positioned with an average interval of about 200μm between the wires. Copper wires distort the coil *RF*-field configuration when they move (*or, when the coil moves relative to the grid*), leading to changes of the oscillator frequency or/and amplitude. The effect is maximum when each wire reaches to the coil center. **Fig.12b** illustrates detected dependence of the oscillator frequency shift, $\delta(\Delta F)$, vs. the lateral position of the metallic-comb relative to the flat-coil face (*relative to "magnetic-field" probe*). Average distance between the detected 6 vertical neighboring peaks on the curve in **Fig.12b** is ~200μm – just in agreement with the experimental setup in **Fig.12a**. That is why, we believe, that *SFCO*-probe may also in future distinguish (*both by amplitude and frequency of the TD-oscillator*) details of the relief of the magnetic or metallic 2D grids, in sub-micrometer scales. For such high lateral resolution, there is need to work out and create the *SFCO* method-based advanced "*magnetic-field*" probe, with a lithographically made single-layer flat coil of about 1mm in diameter [23] – *as an effective needle-type probing instrument with better than* 100nm *predicted lateral resolution.* Such a radically new probe will have considerably *large work-distances* (more than 100μm) *between the probe and surface of the object,* which enables a "visual" control of the local area of probing of the object, and, if needed, application of test perturbations (for example, exposition to laser radiation).

in military science, engineering and Intelligence: to detect onset and amount of attacking soldiery of enemy arm-forces in the absence of direct visibility, and to reveal and detect low-powered nuclear weapon tests. Besides, to solve the perimeter or/and zone-security problems for the intelligence group(s), as well as for the special mission unit(s).

In a precision sensor industry: for creation of non-contact acceleration sensors (*the pickups*) of *super*-high resolution.

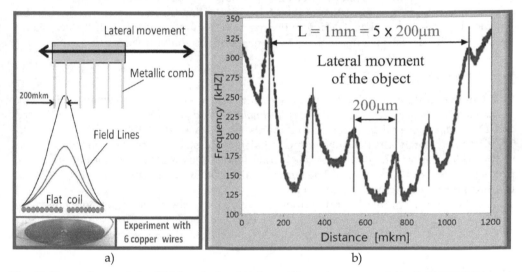

a)   b)

Fig. 12. Dependence of the *SFCO* technique's *TD* oscillator frequency shift $\delta(\Delta F)[kHz]$ (**b**) on the lateral position of the *1D* grid-shaped metallic object (**a**) relative to the "*probe-formative*" flat coil face.

in a basic research: for high-precision measurements of the Casimir Force and very little friction related with it. Besides, high- and/or low-T_c superconductive coil-based 3D analogue of such a new, ultra-sensitive *SFCO*-position sensor seems to be very useful also for the *sub*-Angstrom *spatial*-resolution gravity-wave detection.

4. Conclusion

A new class super-broadband nano-scale-resolution position sensor appeared quite recently. It can be used, in particular, as an additional sensor in seismographs. It enables to extend the frequency-band (up to "zero"), and enhance the absolute-resolution (*sensitivity*) of the vibrometers and seismographs, available on the market, by 10-100 times, depending on the model of the base product (such as the American KS-1/KS-54000 and FBA-23, the European GS-13 and STS-1/STS-2, and the Russian SM-3 – presented and discussed above *SFCO*-sensor was installed just inside the SM-3 seismometer, and compared with its own sensor).

The new position sensor allows transferring of mechanical vibrations of the constructions, buildings & the ground (earth crust) with amplitudes over 1nm, into detectable signal in a frequency range starting practically from quasi-static movements ("zero"!). Such high is the achieved resolution, because due to much higher precision one may measure the frequency of oscillator, compared with the inductance or capacitance of its resonant circuit (even, if use more sensitive AC-bridge technique), oscillators are most suitable sensors for high-precision detection. This is why a very similar position sensor, based on the inductance-change detection of a lithographically made single-layer flat geometry coil, enables three orders less resolution in absolute position sensing [24]. Operation of the new sensor is based on detection of the position changes of a vibrating normal-metallic plate placed near the single-layer flat geometry coil – being used as a pick-up in a stable tunnel diode oscillator. The frequency of the oscillator is used as a detecting parameter in such a sensor, and the measuring effect is determined by a distortion of the MHz-range testing field configuration near the flat coil face by a vibrating plate, leading to the magnetic inductance changes of the coil, with a resolution ~10 pH. This results in changes of measuring oscillator frequency.

5. Acknowledgment

This Chapter was written thanks to study results supported by the Armenian **NFSAT** (*National Foundation of Science and Advanced Technologies*) and the U.S. **CRDF** (*Civilian Research and Development Foundation*) under the Grants ## ISIPA 01-04, GRASP 30/06 and UCEP 07/07. Our research was partially supported also by the state sources of Armenia in frames of the task program on "new materials" (code #041027), as well as in frames of State R&D projects ## 301-0046 and 72-103. The author is also grateful to his staff (YSU, IPR–NAS, and PSI Ltd.) for assistance and useful discussions.
This Chapter was prepared and written also thanks to the information and illustrative material given by the chief executive officer (CEO) of the *Precision Sensors/Instrumentation* (**PSI**) *Ltd.*, Mr. Levon P. Thorose, and by leading specialists of this Company.

6. References

[1] Gevorgyan, S.G.; Movsisyan, A.A.; Movsesyan, G.D.; Shindyan, V.A. & Shirinyan, H.G. (1997). On the Possibility of the Creation of Radically New Type Detectors of Particles & Radiation Based on High-T_c Superconductors. *Modern Physics Letters B*, Vol.11, No.25, pp. 1123-1131.
[2] Gevorgyan, S.G.; Movsesyan, G.D.; Movsisyan, A.A.; Tatoyan, V.T. & Shirinyan, H.G. (1998). Modeling of Tunnel Diode Oscillators & Their Use for Some Low Temperature Investigations. *Review of Scientific Instruments*, Vol.69, No.6, pp. 2550-2560.
[3] Gevorgyan, S.G.; Kiss, T.; Movsisyan, A.A.; Shirinyan, H.G.; Hanayama, Y.; Katsube, H.; Ohyama, T.; Takeo, M.; Matsushita, T. & Funaki, K. (2000). Highly Sensitive Open-Flat Coil Magnetometer for the $\lambda(H,T)$ Measurements in Plate-Like High-T_c Cuprates. *Review of Scientific Instruments*, Vol.71, No.3, pp. 1488-1494.
[4] Gevorgyan, S.G.; Kiss, T.; Ohyama, T.; Movsisyan, A.A.; Shirinyan, H.G.; Gevorgyan, V.S.; Matsushita, T.; Takeo, M. & Funaki, K. (2001). Calibration of the Open-Flat Coil-based Tunnel Diode Oscillator Technique (OFC magnetometer) for Quantitative Extraction of Physical Characteristics of Superconductive State. *Physica C: "Superconductivity and its Applications"*, Vol.366, No.1, pp. 6-12.

[5] Gevorgyan, S.G.; Kiss, T.; Ohyama, T.; Inoue, M.; Movsisyan, A.A.; Shirinyan, H.G.; Gevorgyan, V.S.; Matsushita, T. & Takeo, M. (2001). New Paramagnetic Peculiarity of the Superconductive Transition Detected by a Highly Sensitive OFC magnetometer. *Superconductor Science and Technology*, Vol.14, No.12, pp. 1009-1013.

[6] Gevorgyan, S.G.; Kiss, T.; Shirinyan, H.G.; Movsisyan, A.A.; Ohyama, T.; Inoue, M.; Matsushita, T. & Takeo M. (2001). The Possibility of Detection of Small Absorption in HTS Thin Films by Means of the Highly Sensitive OFC Magnetometer. *Physica C: "Superconductivity and its Applications"*, Vol.363, No.2, pp. 113-118.

[7] Gevorgyan, S.G.; Kiss, T.; Inoue, M.; Movsisyan, A.A.; Shirinyan, H.G.; Harayama, T.; Matsushita, T.; Nishizaki, T.; Kobayashi, N. & Takeo, M. (2002). Peculiarities of the Magnetic Phase Diagram in Small-size Untwinned $YBa_2Cu_3O_y$ Crystal Constructed by Highly Sensitive OFC-magnetometer. *Physica C: "Superconductivity and its Applications"*, Vol.378-381 (P1), pp. 531-536.

[8] Gevorgyan, S.; Shirinyan, H.; Manukyan, A.; Sharoyan, E.; Takeo, M.; Polyanskii, A.; Sarkisyan, A. & Matsushita, T. (2004). Flat Coil-based Tunnel Diode Oscillator Enabling to Detect Real Shape of the Superconductive Transition Curve & Capable of Imaging the Properties of HTSC Films with High Spatial-resolution. *Nuclear Instruments & Methods in Physics Research A* (NIM-A), Vol.520, No.1-3, pp. 314-319.

[9] Gevorgyan, Samvel; Gevorgyan, Vardan. & Karapetyan, Gagik. (2008). A Single-layer Flat-coil-oscillator (SFCO) based Super-broadband Position Sensor for Nano-scale-resolution Seismometry. *Nuclear Instruments & Methods in Physics Research A*, Vol.589, No.3, pp. 487-493.

[10] Gevorgyan, S.G.; Gevorgyan, V.S.; Shirinyan, H.G.; Karapetyan, G.H. & Sarkisyan, A.G. (2007). A Radically New Principle of Operation Seismic Detector of Nano-scale Vibrations. *IEEE Transactions on Applied Supercond.*, Vol.17, No.2, pp. 629-632.

[11] Van Degrift, C.T. (1975). Tunnel Diode Oscillator for 0.001 ppm Measurements at Low Temperatures. *Review of Scientific Instruments*, Vol.46, No.5, pp. 599-607.

[12] Van Degrift, C.T. & Love, D.P. (1981). Modeling of Tunnel Diode oscillators. *Review of Scientific Instruments*, Vol.52, No.5, pp. 712-723.

[13] Bath, M. (1973). Introduction to Seismology, *John Wiley & Sons, NY*.

[14] Sharvin, Y.V. & Gantmakher, V.F. (1960). Dependence of the Superconductive Penetration Depth on the Value of Magnetic Field. Value. *Journal of Experimental and Theoretical Physics (Soviet JETP)*. Vol.39, p. 1242.

[15] Gevorgyan, S.; Kiss, T.; Movsisyan, A.; Shirinyan, H.; Ohyama, T.; Takeo, M.; Matsushita, T. & Funaki, K. (2000). Advantages of λ Measurement in Flat Geometry High-T_c Cuprates by an Open-Flat Coil Magnetometer Demonstrating Its Wide Possibilities For Detection. *Nuclear Instruments & Methods in Physics Research A* (NIM-A), Vol.444, No.1-2, pp. 471-475.

[16] Giaever, I. (1960). Electron Tunneling Between Two Superconductors. *Physical Review Letters*, Vol.5, No.10, pp. 464-466.

[17] Muradyan S.T., Gevorgyan S.G. (2008) Investigation of TD-oscillators activated on the internal capacitance of their coils. *Journal of Contemporary Physics* (NAS of Armenia), Allerton Press, Inc., Vol.43, No.2, pp. 97-100.

[18] Gevorgyan S.G., Muradyan S.T., Kurghinyan B.K., Qerobyan M.I. (2011) Needle-like Properties of the probing RF field of *"magnetic-field"* probes based on the single-

layer flat coils. *Proceedings of the Yerevan State University*, Physical and Mathematical Sciences, No.3(226), pp. 47-51.

[19] Binning G. and Rohrer H. (1980) "Scanning Tunneling Microscope". *U.S. Patent 4,343,993*, Aug.10, 1982. Field: Sep.12.

[20] Binning G. et al. (1982) *Appl. Phys. Lett.*, Vol.40, p.178.

[21] Binning R., Quate C., et al. (1986), *Phys. Rev. Lett.*, Vol.56, p.930.

[22] Bikov V.A. (2000) "Topical review on Probe Microscopes". *Russian Doctoral Degree Dissertation*, Moscow – in Russian.

[23] Spiral Chip Inductors (U.S. Microwaves production – *www.usmicrowaves.com*).

[24] Roger A. (1996). Coil-based Micromachined Sensor Measures Speed & Position for Automotive Applications. *Electronic Design*, December 16, pp. 34-37.

5

Fibre-Optic Sagnac Interferometer as Seismograph for Direct Monitoring of Rotational Events

Leszek R. Jaroszewicz[1], Zbigniew Krajewski[1] and Krzysztof P. Teisseyre[2]
[1]Military University of Technology
[2]Institute of Geophysics Polish Academy of Sciences
Poland

1. Introduction

The possibility of a direct monitoring of rotational events has an important role in the seismological sciences as well as in the applied physics regarding large engineering structures.

According to the first aspect, a possibility of existence of the rotational phenomena in the seismic field has been discussed from the beginning of the earthquakes investigations. The interest in these phenomena has been stimulated by strange, rotary and even screw-like deformations that occur after earthquakes, often appearing on parts of tombs and monuments (Ferrari, 2006; Kozák 2006). The classical textbooks on seismology deny the possibility that the rotational phenomena, especially in form of seismic rotational waves – SRW, could pass through a rock, so the earthquake rotational phenomena were explained by an interaction of standard seismic waves with a compound structure of objects they penetrate, which, in fact, might be the case (Teisseyre & Kozak, 2003). Nevertheless, it was theoretically proved that even the SRW could propagate through grained rocks; later on, this possibility was extended on rocks with microstructure or defects (Eringen, 1999; Teisseyre & Boratyński, 2002) or even without any internal structure (Teisseyre, 2005; Teisseyre et al., 2005; Teisseyre & Górski, 2009), due to the asymmetric stresses in the medium. It should be noticed that the SRW were for the first time effectively recorded in Poland in 1976 (Droste & Teisseyre, 1976). From this time, waves of this type have been studied in a few centers over the world. Taking into consideration large engineering structures, the rotational events monitoring is connected to the torsional effects in structures as well as to the interstory drift. Since the application of new materials and technologies for building constructions, they have irregular structures in-plane which causes difficulties in designing of the horizontal rotations of these structures especially during earthquakes (Schreiber et al., 2009). Recently in the above areas, the first monographs have been published (Teisseyre et al., 2006, 2008; Lee et al., 2009), covering the theoretical aspects of the rotation motion generation and propagation, as well as the examples of the field experiments.

A further experimental verification of the existing rotational phenomena in seismic events needs a new approach to the construction of the measuring devices, because the conventional seismometers are inertial sensors detecting only linear velocities. Similarly, the measurements of torsional response and interstory drift are reasonably easy on small scale models in a laboratory (Kao, 1998) but are much more difficult in real structures. The first of them can be measured by using a pair of accelerometers and then dividing differences in the horizontal accelerations by the distance between them in a direction perpendicular to the measured motion. Then this has to be integrated twice with respect to the time needed to give the torsional rotations (Schreiber et al., 2009). However, the inherent sensor drift and the small offset from zero in the absence of an input signals are the important limitations of this technique. According to the measurement of interstory drifts, it is, in principle, possible to arrange a frame from the floor below to near the ceiling above to set up the displacement transducers to measure the difference in displacements (McGinnis, 2004). However, again far from the hardware complexity of this approach, it is also vulnerable to building deformations.

For the above reason, the new instrumentations are important, especially those designed for an investigation of very small rotations. The near–field studies for the understanding of the mechanics of earthquakes is extensively reviewed by Kamamori (1994) and can be summarized as requirement for instruments with frequency range below 100 Hz and resolution in range of 10^{-6}-10^{-9} rad/s/Hz$^{1/2}$ for the SRW. Whereas the engineering strong-motion seismology needs devices operating in a frequency range of 0.05 – 100 Hz with resolution 10^{-1} – 10^{-6} rad/s/Hz$^{1/2}$ (Cowsik et al., 2009).

Since the Sagnac effect (Sagnac, 1913) measures the rotation directly, an application of the sensor based on this effect seems to be ideal for the construction of the rotational seismometer - RS. Its greatest strength is the fact that it does measure absolute rotations or oscillations, so that it does not require any external reference frame for its measurement. This means that it measures true rotations even during an earthquake, where nothing remains static. Since it is an entirely optical device, it does not have the problems that characterize inertial mass transducers, also. We distinguish two systems based on the Sagnac interferometer: a ring laser rotational seismometer - RLRS (Schreiber et al., 2001), and a fibre-optic rotational seismometer - FORS (Takeo et al., 2002; Franco-Anaya et al., 2008; Schreiber et al., 2009). The first of them is a stationary system constructed for investigation disturbances in the Earth rotation, whereas the second one based on the application commercially available fibre-optic gyroscope - FOG.

Even though 40-ty years of the FOG investigation gives very precise systems useful for different areas including inertial navigation, their constructions are optimized for the detection of angular changes rather than rotation speed, then may generate the same difficulties during the investigation rotational phenomena. For the above reason in this chapter we conclude ours experiments connected to another approach to the rotation phenomena investigation (Jaroszewicz et al., 2003). We started the second part with a short description of the Sagnac effect with the same historical review of its application as FOG. In part 3, which is the main part of this chapter, we described the Autonomous Fibre-Optic Rotational Seismograph – AFORS, with its optical part based on the FOG construction, whereas the special autonomous signal processing unit – ASPU optimizes its operation for the measurement of rotation speed instead of angular changes. Finally, in part 4 we presented the same results obtained during the application of these systems for the rotational events investigation as well as a monitoring of building torsional moves.

2. Sagnac effect and its application as FOG

Sagnac (1913) first demonstared the feasibility of an optical experiment capable of indicating the state of rotation of the frame of reference in which his interferometer was at rest. The basic principle of Sagnac's interferometer is given in Fig. 1a. The input light beam is split by a beam splitter into a beam circulating in the loop in a clockwise - cw direction and a beam circulating in the same loop in a counterclockwise - ccw direction. The two beams are reunited at a beam splier so that interference fringes are observed in the output light. When the whole interferometer with a light source and the fringe detector is set in rotation with an angular rate of Ω rad/s, a fringe shift ΔZ with respect to the fringe position for stationary interferometer is observed, which is given by the formula:

$$\Delta Z = \Omega \cdot A / \lambda_0 c, \tag{1}$$

in which A is the area enclosed by the light path. The vacuum wavelength is λ_0 and the free-space velocity of light is c. The scalar product $\Omega \cdot A$ denotes that ΔZ is proportional to the cosine of angle between the axis of rotation and the normal to the optical circuit. Sagnac also established that the effect does not depend on the shape of the loop or the center of rotation.

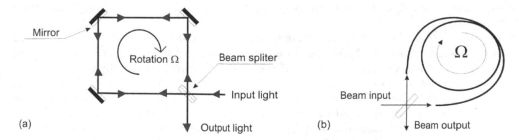

Fig. 1. Schematic of Sagnac's interferometer (a) and its implementation in fibre optic technique (b)

It should be noticed that a German graduate student, Harress (1912), performed a very similar experiment for a thesis project a few years before Sagnac did his experiment. Harress used an optical circuit which consisted of a ring of total reflecting prism, but his objective was quite different from Sagnac's. According to Sagnac's data (Sagnac, 1914), for the wavelength of indigo mercury light and a loop area A=866 cm², a fringe shift of 0.07 fringes was clearly detectable for the rate of rotation of 2 rps. However, according to the other data (Post, 1967) the fringe shift detectability at that time was probably of an order of 0.01 of a fringe, so precision of the Sagnac's experiment therefore may have been close to marginal.

A Sagnac experiment of great precision was subsequently performed by Pogany (1926). With a loop area A=1178 cm², Ω=157.43 rad/s, and λ_0=546 nm, he reproduced within 2% the theoretically expected fringe shift ΔZ=0.906. Michelson and Gale (1925) succeeded in demonstrating the rotation of the Earth by means of the Sagnac effect, also. To obtain the required sensitivity they had to choose an unusually large size (rectangular 0.4 x 0.2 mile) for the surface area enclosed by the beam. Summarizing, the experiments of Sagnac, Pogany, and Michelson-Gale and the results of Harress, as reinterpreted by Harzer (1914), the following features of the Sagnac effect according to the fringe shift can be given (Post, 1967):

a.   obeys formula (1),
b.   does not depend on the shape of surface area A,
c.   does not depend on the location of the center of rotation,
d.   does not depend on the presence of a commoving refracting medium in the path of the beam.

The fibre-optic version of the Sagnac interferometer uses a long length optical fibre L coiled in a loop of diameter D, as it is shown in Fig. 1b (Vali & Shorthil, 1976). In this approach, instead of the fringe shift ΔZ, a phase shift $\Delta\phi$ is produced between cw and ccw propagating light, given by

$$\Delta\phi = \frac{2\pi LD}{\lambda_0 c_0}\Omega \qquad\qquad (2)$$

where Ω is the rotation component in the axis perpendicular to the fibre-optic loop. In other words, the sensitivity of the Sagnac interferometer in this approach is enhanced not only by increasing the physical sensor loop diameter but also by increasing the totals length of the used fibre. It is easy to see that three such interferometer with loops plane in perpendicular directions give information about a space vector of the rotation rate. This data after an integration in time domain shows the position changes in space – and it is idea of the optical gyroscope.

35-years after the above date its technical application as the FOG is the best recognized interferometric sensor made in the fibre-optic technology. However, because its useful signal is the angular changes, the detected phase shift $\Delta\phi$ is integrated in time needed to give it. Moreover, for a desired rotation rate in the range of 10^{-6} – 10^{-9} rad/s the Sagnac effect generates a very small phase shift, so needs to be separated from other disturbances and protected so that the Sagnac effect is the unique nonreciprocal effect in the device. For the above reason all FOG uses shown in Fig. 2 the reciprocal configuration (Urlich, 1980) also called the minimum configuration (Arditty & Lefevre, 1981) where a perfect balance between both counter-propagating paths is obtained simply with a truly single-mode (single spatial mode and single polarization) filter at the common input-output port of the interferometer, even if the propagation is not single-mode along the rest of the interferometer.

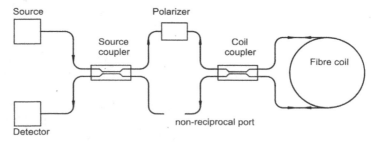

Fig. 2. The minimum configuration of the FOG

The FOG interferometer using the minimum reciprocal configuration yields a raised cosine response as any interferometer. It is classical to bias such an even response that has a maximum at zero by modulating the abscissa parameter and demodulating the output signal, which yields an odd response. The FOG uses a reciprocal phase modulator - PM at the end of the coil, which yields, because of the propagation delay, as a modulation of the phase difference without any residual zero offset (Martin & Winkler, 1978). This was a very important step in the progress of performance, but it was not enough for an ultimate performance which is obtained only if the unbiased response is perfectly even and the biasing modulation has only odd frequencies. For the above reason, the PM being actually a delay line filter, the operation at the so-called proper or eigen frequency (Bergh et al., 1981) - the delay through the coil is half the period of modulation, suppresses the residual even harmonics which are always present because of spurious nonlinearities of the modulation chain. Today the FOG system using also a broadband source has the intensity statistics that happens to cancel the Kerr effect induced phase difference in a Sagnac interferometer (Ezekiel et al., 1982). Such a broadband source is also needed, as it is well-known today, to remove coherence related noise and drift due to backscattering and backreflection as well as lack of rejection of the polarizer (Fredricks & Ulrich, 1984; Lefèvre et al., 1985a; Burns, 1986). Finally, for achieved the high scale factor linearization, FOG utilizes a digital phase step feedback (Lefèvre et al., 1985b) using the same reciprocal PM as the biasing modulation and an all-digital processing scheme where the gyro modulated signal is sampled with an AD convertor and the demodulation performed by digital subtraction (Auch, 1986; Arditty et al., 1989).

Since all-digital processing scheme implemented in the current FOG system is optimized for the presentation of angular changes rather than rotation rate, the same problems exist for its optimized application for measurement of the last phenomena which are interesting for the rotational seismology. For the above reason in the next part of this chapter we have described our experiments in development of the fibre-optic rotational seismograph system. Its construction is based on experiences according to the FOG development described above, but the system is optimized for a direct measurement of the rotation rate only (Jaroszewicz et al., 2003). Such an approach gives a system which through a direct use of the Sagnac effect can limit drift influence on a device operation. Moreover, the special construction of a signal processing unit protects easily its monitoring via Internet including data collecting and managing as well as device remote control.

3. Autonomous Fibre-Optic Rotational Seismograph

A detailed description of the AFORS system was published previously (Jaroszewicz at al., 2011a, 2011b) hence here we summarized the above data regarding its construction, calibration and management. Now we present two examples of these devices - AFORS-1 located in the Książ (Poland) seismological laboratory for the investigation of the rotational events connected to earthquakes, and AFORS-2 located in Warsaw (Poland) used for initial works connected to the investigation of the irregular engineering construction torsional response and the interstory drift (Jaroszewicz et al., 2011c). Before the end of 2011 the next system AFORS-3 will be available as the replacement to the older version FORS-II mounted in the Ojców (Poland) seismological laboratory (Jaroszewicz & Krajewski, 2008).

The optical head of the constructed AFORS devices uses a fibre interferometer in a minimum optical gyro configuration (Jaroszewicz et al. 2006a), as it is shown in the upper part of Fig. 3a.

The application of the broadband low coherence superluminescent diode – SLD (*EXALOS* - Switzerland with optical power – 20.87 mW, operation wavelength – 1326.9 nm, and spectral radiation band – 31.2 nm,) gives possibility for a minimisation of polarization influence on

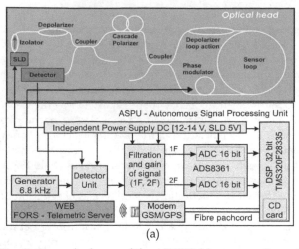

(a)   (b)

Fig. 3. General schema of the AFORS (a): upper – the optical head (generation of the Sagnac phase shift proportional to measured rotation rate Ω), bottom – Autonomous Signal Processing Unit (rotation calculation and recording), (b): general view of all AFORS (top) and ASPU (bottom)

the system operation by achieving light depolarization in a sensor loop (Krajewski et al., 2005). Next the set of cascade fibre polarizers (with total extinction above 100 dB) enables a true single mode operation of the whole system and guarantees that the only nonreciprocal effect in system is the Sagnac effect. Moreover, a 0.63 m diameter sensor loop has been made from a special composite material with permalloy particles for shielding from the magnetic field. A long length of SMF-28 fibre has been winded in a double-quadrupole mode (Dai et al., 2002) with a 0.2 mm Teflon insulation between each fibre layers which is for the thermal stabilization of the sensor's work, for expected 2-4 degree per day temperature fluctuation in seismic observatories. The system optimization made for AFORSs (15000 m length of fibre with attenuation equal to 0.436 dB/km in sensor loop for AFORS-1 and respectively, 15056 m and 0.450 dB/km for AFORS-2) allows for theoretical sensitivity in quantum noise limit (Jaroszewicz & Wiszniowski, 2008) equal to $1.97 \cdot 10^{-9}$ rad/s/Hz$^{1/2}$ and $2.46 \cdot 10^{-9}$ rad/s/Hz$^{1/2}$, respectively for AFORS-1 and AFORS-2. The above mentioned difference between two constructed devices is connected to their total optical loss which is equal to 13.33 dB and 14.47, respectively for AFORS-1 and AFORS-2 (Jaroszewicz et al., 2011b).

The optimisation for a detection of rotation rate is made on the basis of special detection units and utilizes a synchronous detection with properly chosen PM that operates according to the principles presented in part 2 (Krajewski, 2005). For AFORS, a new Autonomous Signal Processing Unit - ASPU (*ELPROMA Ltd*), according to the scheme shown in the lower part of Fig. 1a, has been developped. The ASPU enables the detection of a rotation rate Ω from proper selection (special low-pass filters) and processing (in digital form) the first $A_{1\omega}$ and the second $A_{2\omega}$ amplitude of the harmonic output signal, on the basis of the following relation (Jaroszewicz et al., 2011a):

$$\Omega = S_o \arctan\left[S_e \cdot u(t)\right]; \quad S_0 = (\lambda \bullet c)/(2 \bullet \pi \bullet L \bullet D), \quad u(t) = A_{1\omega}/A_{2\omega} \tag{3}$$

where: S_o – optical and S_e – electronic constants, related to the parameters of used optical and electronic components. The digital form of a signal processing enables the application of the 32-bit signal processor TMS320F283535 (*Texas Instruments*) working with a frequency of 150 MHz as an optimal DSP unit, for a calculation and monitoring of the rotation rate Ω on the basis of signal frames having 1024 length of 16-bit samples. Finally, the obtained results are stored on a CD card and transmitted by a GSM/GPS module to a special WEB FORS - Telemetric Server.

The evaluation of the optical and electronic constans needs a sensor calibration process, which is based on the measurement of the Earth rotation for Warsaw, Poland i.e. Ω_E = 9.18 deg/h $\cong$ 4.45·10⁻⁵ rad/s (Krajewski et al., 2005; Jaroszewicz et al., 2011a). During the calibration the AFORS is mounted vertically on a rotation table. For the sensor loop directed in the East-West direction, the measured rotation equals zero because in this direction, its plane is collinear with the Earth rotation axis, whereas for the North-South direction, the measured signal obtains the maximum plus or minus values because the plane of a sensor loop is perpendicular to the vector component of the Earth rotation Ω_E. For two existing systems we obtained the following values for the above constants: S_0= 0.0043 s⁻¹, S_e=0.0144 for AFORS-1 and S_0=0.059 s⁻¹, S_e=0.0134 for AFORS-2.

After the practical construction of the AFORS devices, their accuracy has been checked. However, this work made in MUT located in Warsaw city, could give limited information on the system accuracy because of urban noises. Figure 4 summarizes these measurements. Since the ASPU allows for step changes of the detection frequency band in the range from 0.83 Hz to 106.15 Hz (Jaroszewicz et al., 2011b), the obtained accuracy is at the level of 5.07·10⁻⁹/4.81·10⁻⁹ rad/s - 5,51·10⁻⁸/6.11·10⁻⁸ rad/s (for AFORS-1/AFORS-2), respectively for the lower and higher working frequency band. As one can see, the obtained values are well correlated with Ω_{min} in quantum noise limitation. It should be noticed that the linear dependence of AFORS sensitivity and accuracy in the detection frequency range is the advantage of this system, taking into account the expected frequency characteristics of the rotational seismic waves (Teisseyre et al., 2006). For comparison, Fig. 4 includes also the measured accuracy of the older system FORS-II; which was 4.3·10⁻⁸ rad/s (Jaroszewicz et al., 2006; Jaroszewicz & Krajewski, 2008), at 20 Hz - the fixed detection frequency band.

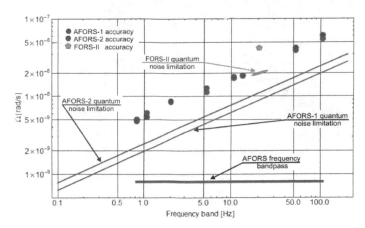

Fig. 4. The accuracy measured in Warsaw, Poland for the chosen detection band for three devices: AFORS-1, AFORS-2 and FORS-II

A *FORS-Telemetric Server* with its main page shown in Fig. 5a (reader can use http://fors.m2s.pl with login and password - AFORSbook for free access to the system) is used for data storing and for monitoring the work of the FORS-II and the AFORSs. Because ASPU of the AFORS contains a GSM/GPS module and an independent power supply for all electronic components of the system, hence the AFORS is fully autonomous and mobile system. In this moment, the 3 devices are managed via server: FORS-II, AFORS-1, AFORS-2 located in Ojców, Książ and Warsaw (all in Poland), respectively (Jaroszewicz et al., 2011b) as it is shown in Fig. 5b.

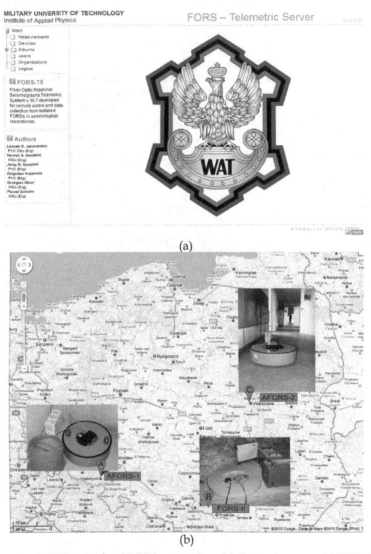

(a)

(b)

Fig. 5. Elements of WEB page for AFORS managing: (a) the main page of *FORS - Telemetric Server*, (b) the GOOGLE map with devices localization

The applied technology gives possibility for the remote (via Internet) controlling and changing of all electronic parameters of the ASPU for a given sensor made according to the AFORS technology, as presents, for example, the bookmark *Config* for the AFORS-1 in Fig. 6a. This remote control may comprise a software upgrade. Moreover, the bookmark *Data&Variables* (Fig. 6b) monitors, for given AFORS, in real time the main data and variables with possibility for the remote changing of the threshold – the level of signals which initialize automatic data storing and its GSM transfer. Additionally, the top right corner of bookmarks for the given system on server contains the information on a current date and time and the four main AFORS's parts of state of work (good – as green, partially good as

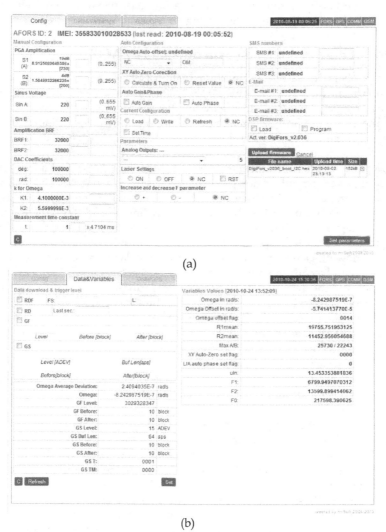

(a)

(b)

Fig. 6. The view of two main bookmarks for AFORS-1 at the *FORS - Telemetric Server*: (a) *Config*, and (b) *Data&Variables*

yellow or no work – as black, respectively). The bookmark *GSM/GPS* (not shown in figure, see for example Jaroszewicz et al., 2011a) monitors in real time the GSM parameters as well as the GPS parameters which include the AFORSes' global localization (see Fig. 5b). Yet another bookmark named *Measurement* presents the collection of data recorded by different devices connected to the server. These data are stored with the main parameters of AFORS in the recording time: ADEV – rotation rate average deviation in rad/s, Omega Offset – rotation rate offset in rad/s, GS Level/Before/After - adjusted level of signal for data stored, and ΔB - adjusted detection band. In this way, in our opinion, the AFORSes with their management via *FORS - Telemetric Server* are fully adopted for monitoring of rotational phenomena connected to earthquakes as well as torsional response and interstory drift of the irregular structures in-plane existing during any ground moves.

4. Examples of the experimental data obtained by AFORS

The previously obtained data from an older system FORS-II have been wide discussed as well as summarized (Jaroszewicz et al., 2006, 2008). For the above reason here we present the summarized data obtained with regard to the AFORS application where AFORS-1 is installed in the Książ (Poland) seismological laboratory for the recording of the rotational phenomena connected to earthquakes, whereas AFORS-2 has been used in the initial experiments for monitoring building rotational moving.

The main source of disturbance during an investigation by AFORS-2 of a building rotation moving was an urban ground motion generated by tram moves within a 50 m distance from a building wall parallel to it. The investigated building is a light construction (five floors of aluminium structure with sandwich walls and ceilings), and the AFORS-2 has been installed, subsequently on the second and first floors in the hall, in the same vertical position (with accuracy of about 10 cm). Since it is an old building with asbestos used as an inner wall isolation, now it is not in use by the academy anymore, so we expected that the recorder signals will be connected to an external perturbation. Figure 7 presents the building moves recorded on the first and the second floors (difference about 3 m of height) for relatively the same ground motion generated by tram moves nearly by midnight on July 13th (AFORS-2 on the first floor) and July 14th (AFORS-2 on the second floor). Since it was a middle of the night during summer holidays, the academy area was empty which had a direct influence on recorded signals and they were very clear. As one can see in the above experiment the accuracy for the AFORS-2 was $3.15 \cdot 10^{-6}$ rad/s and $7.91 \cdot 10^{-6}$ rad/s (see ADEV parameter in the left down corner of two pictures in Fig. 7), for the chosen detection band equal to 21.23 Hz. The amplitude of the detected rotation rate was about twice higher for the second floor, and was much higher than the system accuracy (more than ten orders).

The urban noise influence on the recorded signals can be observed on the data presented in Fig. 8, which have been obtained in the morning when the Academy opened for work. As one can see the higher amplitude as well as frequency were observed in this time. However, again the much higher amplitude of torsional moves of the building is observed on the higher floor of building.

The above initial results show that the device type AFORS can be useful for a continuous monitoring of an engineering structure of, for example, multi-storey buildings with regard to the investigation of their torsional rotations as well as measuring interstory drifts. These measurements are made without any reference frame which is very important during earthquakes and may be made only by a system based on the Sagnac effect. In comparison to the commercially available FOG instruments such as μFORS-1, the proposed system is designed for a direct measurement of a rotational rate, whereas any FOG measures change the angle which is written in their inner electronic system and difficult to direct changes. Additionally, our system prepared according to the AFORS technology has developed the software designed for the Internet system monitoring as well as the remote control which can manage a large numbers of such devices in a useful way for the operator.

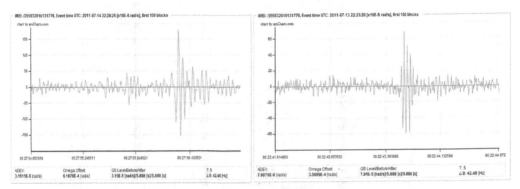

Fig. 7. The data recorded on second (left) and first (right) floor as response for ground moves after tram pass through street

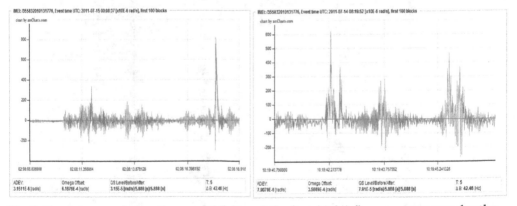

Fig. 8. The data recorded on the second (left) and the first (right) floor as a response for the ground moves generated by the street morning intensity within a distance of about 50 m from and parallel to the long building wall

At beginning of July, 2010 the AFORS-1 has been installed in the Książ (Poland) seismological observatory together with a set of the Two Antiparallel Pendulum Seismometers (*TAPS-1* and *TAPS-2*) constructed by the Institute of Geophysics (Teisseyre et al., 2003). As usually, TAPSes are placed perpendicularly, in directions N-S and E-W.

As the first example we repeat here (Fig. 9) the histogram of the previously analyzed data (Jaroszewicz et al., 2011b) collected in Książ on March 11th, 2011 at 6 h 58 min. (after Honshu earthquake, M=9.0 on 11 March 2011 at 5 h 46 min. 23 s, recorded in Książ on 11 March 2011 at 5 h 58 min. 35 s.). The above data were obtained from the common for AFORS

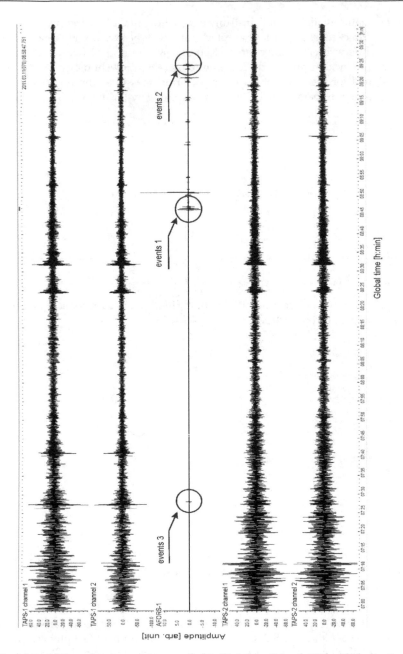

Fig. 9. The plots of the seismic events recorded in Książ on March 11th, 2011, starting from 6 h 58 min, after the Honshu M=9.0 earthquake, all times UTC (Jaroszewicz et al., 2011b)

and TAPSes standard seismic recording system named KSPROT with the samples of a signal with the frequency 12.8 kHz and after re-sampling stores with the frequency 100 Hz.

It should be noticed that in the previously paper (Jaroszewicz et al., 2006), we used a wrong name for this station, KST. We underline that here only the AFORS-1 shows the rotational component in a direct way (plots marked as AFORS-1). The rotational component is obtained also from the TAPS system, calculated from the recordings of linear motions in four channels (data named TAPS-1 channel 1, TAPS-1 channel 2, TAPS-2 channel 1 and TAPS-2 Channel 2) with the application of a suitable mathematical procedure, which has been widely described in the previous paper (Solarz et al., 2004). Since AFORS records rotation in a direct way, we use this recording as the reference source, despite that the rotations calculated from TAPSes are generally poorly correlated with it. In the results presented in Fig. 9, a good correlation has been found mainly in short bursts of seismic oscillations, marked here as "events 3", "events 1" and "events 2". These were clusters of short peaks, found when the traces of the great Honshu Earthquake were studied (without much success in the domain of rotational field component, but this is not strange concerning the distance to the earthquake focus and the characteristics of our instruments). After the preliminary analysis, from each group only one event was chosen, and these we call event 3, event 1 and event 2 accordingly.

For event 1, the data from AFORS-1 were identified by the FORS-Telemetric Server (see Fig. 10a) as the rotational event of an amplitude of about $15 \cdot 10^{-6}$ rad/s with AFORS-1 accuracy equal to $4 \cdot 10^{-8}$ rad/s for the given frequency bandpass which was about 10.6 Hz. The rotation calculated from the linear motions recorded in TAPS system has similar characteristic, as is seen in Fig. 10b (with arbitrary units of amplitude). Nevertheless, we observed some time advance in a relation to the AFORS-1 registration as well as apparent disturbances, visible before and after the event; these may result from a limited accuracy of TAPSes as was previously mentioned (Jaroszewicz & Krajewski, 2008).

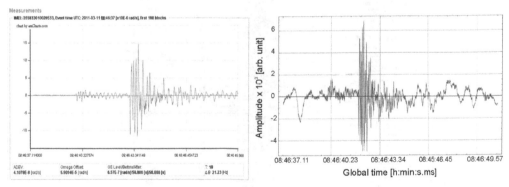

Fig. 10. The rotational event 1 from Fig. 5 on 11 March 2011 – a): recorded from AFORS-1 in the FORS -Telemetric Server at 8 h 46 min; – b): calculated from TAPSes four channels (Jaroszewicz et al., 2011b)

For further study, the recordings of several mining shocks have been chosen; these shocks occurred in the Legnica-Głogów Copper Mining District - LGOM in Western Poland. Some typical results of the analysis of strong shocks are presented here; for the event which occurred in April 30th at 03:32 UTC - the magnitude 2.9 had been found, for the event from June 28th, 23:16 UTC – magnitude 3.2. For viewing the data, for the normalization of the sampling to the first channel and for writing selected time-periods as ASCII, we used the programs written by dr Jan Wiszniowski of the Institute of Geophysics, P.A.Sci. Subsequent analysis was done in Matlab®.

The rotational motions in the trace of a seismic event, recorded from AFORS-1 and indirectly obtained from TAPSes, differ substantially, as it is seen in Fig. 11. The signal obtained from AFORS-1, here – the middle plot, has much more peaks and indentations than both the channel 1 and channel 4, which are two of the four channels of the TAPSes system. The signals from TAPSes are shown in μm/s, while the rotational signal from AFORS-1 is in conventional units. In the following Figures, the rotations found from TAPSes system are plotted in rad/s.

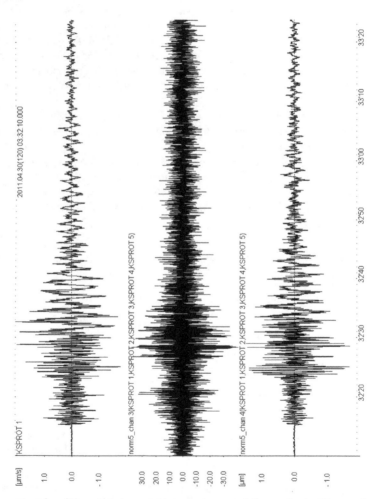

Fig. 11. The example of the seismic event traces, prepared for analysis. Upper channel – channel 1, first of two from the first electromechanical rotational seismometer TAPS. Middle channel – rotational seismogram from AFORS-1, sampling-normalized to the channel 1. Lower channel – first from the second TAPS, sampling-normalized to the channel 1. The mining seismic event in LGOM (western Poland) on 2011.04.30, at 03:32 UTC

For the analysis, we divided each signal into several frequency bands, then searched for similar – and dissimilar! – rotational motions. As it is seen in the Fig. 12, even this method did not reveal great similarity between the rotations obtained from AFORS-1 and from TAPSes. Here, the time-period when the P waves have arrived is shown. For all the transformations, we applied the same procedures, including the digital filters which we constructed in Matlab®.

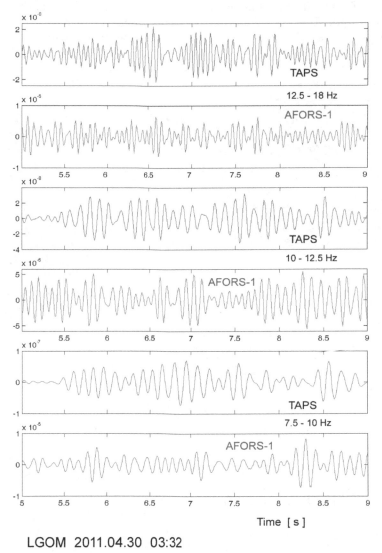

LGOM 2011.04.30 03:32

Fig. 12. Rotation motions in the recording of a mining seismic event, during P waves arrivals, as seen in three frequency bands: 12.5-18 Hz, 10-12.5 Hz and 7.5-10 Hz. In each of the paired diagrams, the upper one shows the rotation calculated from the system of TAPSes, the lower one – the rotation sensed by the AFORS-1

Further, we tried to compare both rotational signals after averaging each of them for consecutive time-periods of 0.5 s, that is – 50 samples. The results, for six chosen frequency bands, are shown in Figs. 13 – 16. The averages of the absolute values of rotations are plotted as thick line. Additionally, two other averages are added. The thin continuous is for acceleration (in fact – the difference between neighbouring samples) – again, the absolute values are averaged for consecutive stages. Finally, thin dashed line shows the average of the squared rotational signal. These additional plots were normalized to the rotational signals, so their maxima coincide. The Figs. 13 and 14 are obtained from the analysis of the shock from April 30th; the Figs. 15 and 16 – from the analysis of the stronger shock, that of June 28th.

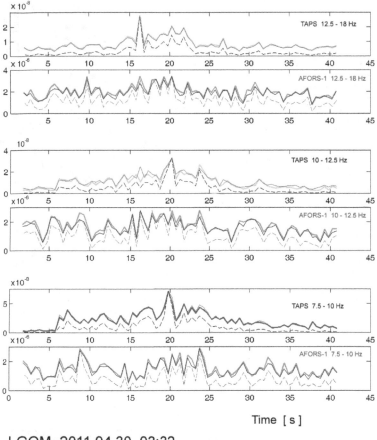

Time [s]

LGOM 2011.04.30 03:32

Fig. 13. The rotations, their squares and time-differentials equivalent to rotational acceleration, averaged in consecutive stages, 50 or 0.5 second long. Before averaging, the rotations and their differentials were transformed into the absolute values. The seismic event in LGOM, 2011.04.30, at 03:32 UTC. The upper frequency bands: 12.5-18 Hz, 10-12.5 Hz and 7.5-10 Hz. In each of the paired diagrams, the upper one shows signals obtained from the system of TAPSes, the lower one – from the AFORS-1

Searching for a similarity between peaks, indentations and their sequences, placed in the same points in the both plots of rotations' average values, is the searching for hidden similarities between the rotations obtained from electromechanical seismometers system and these obtained from AFORS-1. As it is seen in the Figs. 13 – 16, only partial similarity may be found, in the specific frequency bands only. In the two cases presented here, these were the frequency bands 3 - 4.8 Hz and 10 - 12.5 Hz, and to lesser extent – 12.5 - 18 Hz.

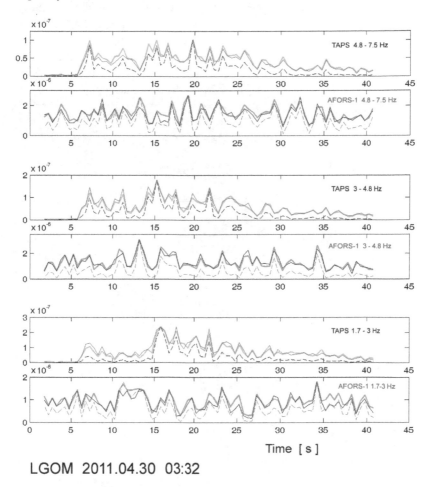

LGOM 2011.04.30 03:32

Fig. 14. Analysis as in Fig. 13; the same seismic event. The lower frequency bands: 4.8-7.5 Hz, 3-4.8 Hz and 1.7-3 Hz

It is often hard to find the beginning of the seismic event's trace in the diagram of rotational motions; dividing the signals into frequency bands usually helps in the case of recordings from electromechanical seismometers (TAPSes), but for recordings from AFORS-1 the improvement is smaller. For the frequencies in the band 12.5 - 18 Hz, and higher, the rotational trace of a mining seismic event is often obscured by the noise. This is more clearly seen in the readings from AFORS-1 because of a high sensitivity of this equipment to high frequency rotations. The mentioned noise might be of an instrumental or external origin, the latter is more probable. Nevertheless, the rotational traces of the seismic events in LGOM area also bear high frequency rotations, in various portions. These high frequency rotations may originate in the focus, or in the vicinity of the seismic station, as a response to the arriving seismic waves. We relate the latter explanation also to the short bursts of high frequency rotational motions, which quite often accompany the recording of an earthquake or other seismic event (Jaroszewicz et al., 2011b).

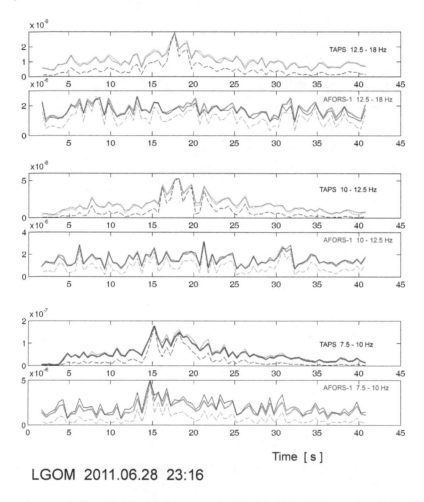

Time [s]

LGOM 2011.06.28 23:16

Fig. 15. Analysis as in Fig. 13. The seismic event in LGOM, 2011.06.28, at 23:16 UTC. The upper frequency bands: 12.5-18 Hz, 10-12.5 Hz and 7.5-10 Hz

When a recording from AFORS-1 is analyzed in low frequencies, below 1.7 Hz, a trace of a seismic event is almost not visible, and it is barely discernible also in the frequency band 1.7 – 3 Hz. That's unfortunate, as a substantial part of rotational oscillations comes in low frequencies. The share of low frequency rotations varies between local events, as was found for several shocks recorded in Ojców, Poland (K. P. Teisseyre, 2006) and in l'Aquila, central Italy (K. P. Teisseyre, 2007). In each of these studies, the rotational motions were derived from the set of two TAPSes, in other words – from two pairs of electromechanical horizontal seismometers.

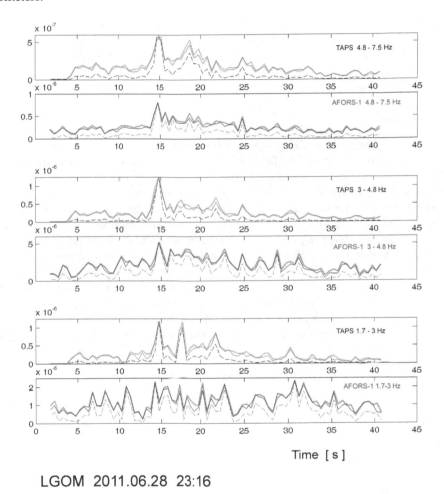

Time [s]

LGOM 2011.06.28 23:16

Fig. 16. Analysis as in Fig. 13. The seismic event in LGOM, 2011.06.28, at 23:16 UTC. The lower frequency bands: 4.8-7.5 Hz, 3-4.8 Hz and 1.7-3 Hz

The agreement between the plots of the stage-averaged rotation and the stage-averaged squared rotation is not surprising; it shows that in the seismic energy received at the station, higher amplitudes motions are more important than the smaller ones. On the other hand, in several time-periods (stages), a disagreement between rotational acceleration and the appropriate rotational signal is often found. When the acceleration plot is below that of the adequate rotational velocity plot - see Fig. 14 - this may be explained by a shift from symmetry of average signal amplitudes in a relation to the zero level. But when we take into account that the acceleration has been amplitude-normalized to the rotation, it is clear that we cannot estimate the size of this shift. Such shifts may be present always, or they may be absent only in some cases and frequency bands, for example - this depicted in Fig. 15, two lowermost panels; here – the acceleration coincide with the rotation or is plotted higher except for the summit.

Some methods of cleaning the signals from possible contaminations are under development, and we are going to record rotational motions with the use of two Sagnac-type interferometers placed at some distance. Nevertheless, the studies outlined here show that the field experiments with AFORS-1 bring an important information concerning the seismic field of mining seismic events.

5. Conclusion

The direct measurement of the absolute rotational components with linear changes of the accuracy form $5.1 \cdot 10^{-9}$ to $5.5 \cdot 10^{-8}$ rad/s, for the detection frequencies range from 0.83 to 106.15 Hz is the main advantage of the presented system AFORS. As shown in the first field test, the system type AFORS appears as the promising device to study both the seismic field generated by earthquakes at least at local distance range, and the reaction of buildings and other constructions to strong motions. In the first domain however, the whole detecting/registering system needs further studies – especially, the recordings analysed both in the low frequencies, below 3 Hz and in the relatively high frequencies, from 18 Hz upwards seem to contain too much noise. But most probably, at least part of this noise comes to the equipment from outside. Therefore, also the conditions at the Książ observatory should be checked again, despite that for other scientific equipment, as the regular seismic station KSPROT, these corridors (excavated by the slaves under German supervision during the second world war) provide the satisfactory environment.

6. Acknowledgment

This work was done in 2011 under the financial support of the Polish Ministry of Science and Higher Education under Key Project POIG.01.03.01-14-016/08 "New photonic materials and their advanced application" and partially the Military University of Technology statutory activity No PBS-829.

7. References

Arditty, H. J. & Lefèvre, H. C. (1981). Sagnac Effect in a Fiber Gyroscope, *Optic. Lett.*, Vol. 6, No8, (August 1981), pp. 401-403, ISSN 0146-9592

Arditty, H. J., Graindorge, Ph., Lefèvre, H. C., Martin, Ph., Morisse J. & Simonpiétri, P. (1989). Fiber-Optic Gyroscope with All-Digital Processing, *Proceedings of OFS 6/'89*, Paris, Springer-Verlag Proceedings in Physics, Vol. 44, pp. 131-136

Auch, W. (1986). The Fiber-Optic Gyro – a Device for Laboratory Use Only?, *SPIE Proceedings*, Vol.719, pp. 28-34, ISSN 0277-786X

Bergh, R. A., Lefèvre, H. C. & Shaw, H. J. (1981). All-Single-Mode Fiber-Optic Gyroscope with Long-Term Stability, *Optic. Lett.*, Vol. 6, No.10, (October 1981), pp. 502-504, ISSN 0146-9592

Burns, W. K. (1986). Phase-Error Bounds of Fiber Gyro With Polarization-Holding Fiber, *J. Lightwave Tech.*, Vol. LT4, No.1, (January 1986), pp. 8-14, ISSN 0733-8724

Cowsik, R, Madziwa-Nussinov T., Wagoner K., Wiens D., Wysession, M. (2009). Performance Characteristics of a Rotational Seismometer for Near-Field and Engineering Applications, *BSSA*, Vol.99, No.2B, (May 2009), pp. 1181-1189, ISSN 0037-1106

Dai, X., Zhao, X., Cai, B., Yang, G., Zhou, K. & Liu, C. (2002). Quantitative Analysis of the Shupe Reduction in a Fiber-Optic Sagnac Interferometer, *Opt. Eng.*, Vol.41, No.6, (June 2002), pp. 1155-1156, ISSN 0091-3286

Droste, Z. & Teisseyre, R. (1976). Rotational and Displacemental Components of Ground Motion as Deduced from Data of the Azimuth System of Seismograph, *Publs Inst. Geophys. Pol. Acad. Sc.*, Vol.97, pp. 157-167, ISSN 0139-0109

Eringen, A. C. (1999). *Mirocontinuum field theories. Vol. 1 Foundations and Solids*, Springer-Verlag, ISNB 0-387-95275-6, New York

Ezekiel, S., Davis, J. L. & Hellwarth, R. W. (1982). Intensity Dependent Nonreciprocal Phase Shift in a Fiberoptic Gyroscope, *Springer Series in Optical Sciences*, Vol.32, pp. 332-336

Ferrari, G. (2006). Note on the Historical Rotation Seismographs, in: *Earthquake Source Asymmetry, Structural Media and Rotation Effects*, R. Teisseyre, M. Takeo & E. Majewski, (Eds.), 367-376, ISBN 3-540-31336-2, Springer, Berlin

Franco-Anaya, R., Carr, A. J. & Schreiber. K. U. (2008). Qualification of Fibre-Optic Gyroscopes for Civil Engineering Applications, *Proc. of the New Zealand Society of Earthquake Engineering (NZSEE) Conf.*, (available on CD-ROM), Wairakei, New Zealand, 2008

Fredricks, R. J. & Ulrich, R. (1984). Phase-Error Bounds of Fibre Gyro with Imperfect Polariser/Depolarizer, *Electron. Lett.*, Vol.20, No.8, pp. 330-332, ISSN 0013-5194

Harress, F. (1912). Die geschwindigkeit des lichtes in bewegten korper, *Dissertation*, Jena, Germany

Harzer, P. (1914). *Astron, Nachr.*, Vol.198, pp.377-378

Jaroszewicz, L. R. & Krajewski Z. (2008). Application of the Fibre-Optic Rotational Seismometer in Investigation of the Seismic Rotational Waves, *Opto-Electron. Rev.*, Vol.16, No.3, (September 2008), pp. 314-320, ISSN 1230-3402

Jaroszewicz, L. R. & Wiszniowski J. (2008). Measurement of Short-Period Weak Rotation Signals, in: *Physics of Asymmetric Continuum: Extreme and Fracture Processes*, R. Teisseyre, H. Nagahama & E. Majewski, (Eds.), pp.17-47, ISBN 978-3-540-68354-4, Springer, Berlin

Jaroszewicz, L. R., Krajewski, Z. & Teisseyre, K. P. (2011b). Usefulness of AFORS – Autonomous Fibre-Optic Rotational Seismograph for Investigation of Rotational Phenomena, *Journal of Seismology*, Special issue: Rotational Ground Motions, in press, ISSN 1383-4649

Jaroszewicz, L. R., Krajewski, Z. & Teisseyre, K. P. (2011c). The Possibility of a Continuous Monitoring of the Horizontal Buildings' Rotation by the Autonomous Fibre-Optic Rotational Seismograph AFORS Type, *6th European Workshop on the seismic behaviour of Irregular and Complex Structures (6EWICS)*, Haifa, Israel, 12-13 Sept. 2011 – in press

Jaroszewicz, L. R., Krajewski, Z.& Solarz, L. (2006). Absolute Rotation Measurement Based on the Sagnac Effect. in: *Earthquake Source Asymmetry, Structural Media and Rotation Effects*, R. Teisseyre, M. Takeo & E. Majewski E. (Eds), pp.413-438, ISBN 3-540-31336-2, Springer, Berlin

Jaroszewicz, L. R., Krajewski, Z., Kowalski, H., Mazur, G., Zinówko, P. & Kowalski, J. K. (2011a). AFORS Autonomous Fibre-Optic Rotational Seismograph: Design and Application. *Acta Geophys., Vol.* 59, No.3, (March 2011), pp. 578-596, ISSN 0001-5725

Jaroszewicz, L. R., Krajewski, Z., Solarz, L., Marć, P. & Kostrzyński T. (2003). A New Area of the Fiber-Sagnac Interferometer Aapplication, *Proc. Intern. Micro-Opt. Conf. IMOC-2003*, Part II, *Iguazu Falls, Brazil*, 20-23 Sept. 2003, SBMO/IEEE (2003), pp. 661-666

Kanamori, H. (1994). *Annu. Rev. Earth Planet. Sci.*, Vol.22, pp. 207-307

Kao, C. G. (1998). Design and Shaking Table Tests for a Four-storey Miniature Structure Built with Replaceable Plastic Hinges, *Master's Thesis*, University of Canterbury, Australia

Kozák, J. T. (2006). Development of Earthquake Rotational Effect Study, In: *Earthquake Source Asymmetry, Structural Media and Rotation Effects*, R. Teisseyre, M. Takeo & E. Majewski, (Eds.), 3-10, ISBN 3-540-31336-2, Springer, Berlin

Krajewski, Z. (2005). Fiber-Optic Sagnac Interferometer as Device for Rotational Effect Investigation Connected with Seismic Events (in Polish). *Doctor Thesis*, Military University of Technology, Warsaw, Poland

Krajewski, Z., Jaroszewicz, L. R. & Solarz, L. (2005). Optimization of Fiber-Optic Sagnac Interferometer for Detection of Rotational Seismic Events, *SPIE Proceedings*, Vol.5952, pp. 240-248, ISSN 0277-786X

Lee, W. H. K, Celebi, M., Todorovska, M. I. & Igel, H. (2009). *Rotational Seismology and Engineering Applications*, BSSA, Vol.99, No.2B, (May 2009), ISSN 0037-1106

Lefèvre, H. C., Bettini, J. P., Vatoux, S. & Papuchon, M. (1985a). Progress in Optical Fiber Gyroscopes Using Integrated Optics, *Proceedings of AGARD-NATO*, Vol. CPP-383, pp. 9A1-9A13

Lefèvre, H. C., Graindorge, Ph., Arditty, H. J., Vatoux S. & Papuchon, M. (1985b). Double Closed-Loop Hybrid Fiber Gyroscope Using Digital Phase Ramp, *Proceeding of OFS 3/'85*, San Diego, OSA/IEEE, Postdeadline Paper 7

Martin, J. M. & Winkler, J.T. (1978). Fiber-Optic Laser Gyro Signal Detection and Processing Technique, *SPIE Proceedings*, Vol.139, pp. 98-102, ISSN 0277-786X

McGinnis, (2004). Apparatus and method for detecting deflection of a tower, *U.S. Patent application*, No.0107671 A1, (10 June 2004)

Michelson, A. A. & Gale, H. G. (1925). The Effect of the Earth's Rotation on Light Velocity," *Nature*, Vol.115, (18 April 1925), pp.566-566, ISSN 0028-0836

Pogany, P. (1926). Über die Wiederholung des Harress-Sagnacschen Versuches, *Ann. Physik*, Vol. 385, No11, pp.217-231, ISSN 0003-3804

Post, E. J. (1967). Sagnac effect, *Rev. of Modern Physics*, Vo.39, No2, (April 1967), pp.475-493. ISSN 0034-6861

Sagnac, G. (1913). L'ether lumineux demontr par l'effet du vent relatif d'Etherdanus un interferometre en rotation uniforme, *Compte-rendus a l'Academie des Sciences*, Vol.95, pp. 708-710

Schreiber, K. U., Velikoseltsev, A., Carr, A. J. & Franco-Anaya, R. (2009). The application fiber optic gyroscope for the measurement of rotations in structural engineering, *BSSA*, Vol.99, No.2B, (May 2009), pp. 1207-1214, ISSN 0037-1106

Schreiber, U., Schneider, M., Rowe, C.H., Stedman, G.E. & W. Schlüter (2001). Aspects of Ring Lasers as Local Earth Rotation Sensors, Surveys in Geophysics, Vol.22, No.5-6, (September 2001), pp. 603-609, ISSN 0169-3298

Solarz, L., Krajewski, Z. & Jaroszewicz, L. R. (2004). Analysis of Seismic Rotations Detected by Two Antiparallel Seismometers: Spline Function Approximation of Rotation and Displacement Velocities, *Acta Geophys. Pol., Vol.* 52, No.2, (June 2004), pp. 198–217, ISSN 0001-5725

Takeo, M., Ueda, H. & Matzuzawa T. (2002). Development of Hight-Gain Rotational-Motion Seismograph, *Grant 11354004*, Erthquake Research Institute, Univ. of Tokyo, 5-29

Teisseyre, K. P. (2006). Mining Tremors Registered at Ojców and Książ Observatories: Rotation Field Components, *Publs. Inst. Geophys. Pol. Acad. Sc.*, M-29 (395), pp. 77-92

Teisseyre, K. P. (2007). Analysis of a Group of Seismic Events Using Rotational Components, *Acta Geophys.*, Vol.55, No.4, (April 2007), pp. 535-553, ISSN 0001-5725

Teisseyre, R. & Boratyński, W. (2002), Continuum with Self-Rotation Fields: Evolution of Defect Fields and Equations of Motion, *Acta Geophys.*, Vol.50, No.3, (March 2002), pp. 223-229, ISSN 0001-5725

Teisseyre, R. & Gorski, M. (2009), Transport in Fracture Processes: Fragmentation of Defect Fields and Equations of Motion, *Acta Geophys.*, Vol.57, No.5, (May 2009), pp. 583-599, ISSN 0001-5725

Teisseyre, R. & Kozák, J. T. (2003). Considerations on the Seismic Rotation Effects. *Acta Geophys.*, Vol.51, No.3, (March 2003), pp. 243-256, ISSN 0001-5725

Teisseyre, R. (2005). Asymmetric Continuum Mechanics: Deviations from Elasticity and Symmetry, *Acta Geophys.*, Vol.53, No.2, (February 2005), 115-126, ISSN 0001-5725

Teisseyre, R., Białecki, M. & Górski, M. (2005). Degenerated Mechanics in a Homogeneous Continuum: Potentials for Spin and Twist, *Acta Geophys.*, Vol.53, No.3, (March 2005), pp. 219-23. ISSN 0001-5725

Teisseyre, R., Nagahama, H. & Majewski, E. (Eds.) (2008). *Physics of Asymmetric Continuum: Extreme and Fracture Processes. Earthquake Rotation and Solition Waves*, ISBN 978-3-540-68354-4, Springer-Verlag, Berlin-Heidelberg

Teisseyre, R., Suchcicki, J., Teisseyre, K. P. & Palangio, P. (2003). Seismic rotation waves: basic elements of theory and recording, *Annals Geophys.*, Vol.46, No.4, (August 2003), pp. 671-685, ISSN 2037-416X

Teisseyre, R., Takeo, M. & Majewski, E. (Eds.) (2006). *Earthquake Source Asymmetry, Structural Media and Rotation Effects*, ISBN 3-540-31336-2, Springer, Berlin

Ulrich, R. (1980). Fiber-Optic Rotation Sensing With Low Drift, *Optic. Lett.*, Vol. 5, No5, (May 1980), pp. 173-175, ISSN 0146-9592

Vali, V. & Shorthill, R. W. (1976). Fiber Ring Interferometer, *Appl. Optics*, Vol.15, No5, (May 1976), pp.1099-1100, ISSN 0733-8724

Part 2

Earthquakes and Planning

6

Earthquake Induced a Chain Disasters

Guangqi Chen, Yange Li, Yingbin Zhang and Jian Wu
Kyushu University
Japan

1. Introduction

A strong earthquake not only cause directly damage on constructs but also can result in a series of natural disasters such as landslide, debris flow and flooding. These secondary disasters occurs as a chain disasters as shown in Fig. 1. A strong earthquake can induce a large amount of landslides. And then, a large scale landslide can create a landslide dam when its debris fill into and stop a river. The water impounded by a landslide dam may create a dam reservoir (lake). While the dam is being filled, the surrounding groundwater level rises and causes back-flooding (upstream flooding). And because of its rather loose nature and absence of controlled spillway, a landslide dam is easy to fail catastrophically

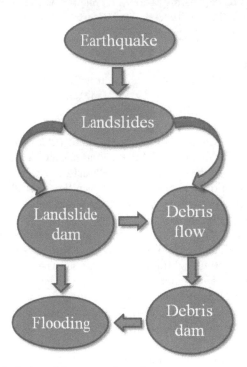

Fig. 1. The disasters chain induced by an earthquake

and lead to debris flow or downstream flooding. Also, since the landslide debris can be removed into a valley or a ravine by excessive precipitation, it is easy to form debris flow, and then create a debris dam sometimes.

In this chapter, we take the 2008 Wenchuan Earthquake (Ms=8.0) as an example to discuss the earthquake induced chain disasters. The characteristics of the earthquake induced landslides are summarized including the discussion of the landslide dams. A landslide susceptibility analysis is carried out and a possible long run-out mechanism is proposed for the study. The characteristics of the debris flows arising from the earthquake are summarized. An approach of simulating debris flow is proposed for predicting the movement behaviours of potential debris flow arising from earthquake. A practical simulation is carried out for verifying the approach.

2. The earthquake induced landslide

A strong earthquake can induce a large amount of landslides and cause very serious property damage and human casualties. This phenomenon was recorded in ancient China dated back to 1789 BCE, and in ancient Greece 2373 years ago (Keefer, 2002). There have been many reports about the very serious damages caused by the earthquake induced landslides for the last few decades. For example, 9,272 landslides induced by the 1999 Chi-Chi earthquake (Ms=7.6) caused 2,400 deaths, more than 8000 casualties and over 10 billion US$ of economical loss in Taiwan(Chang et al. 2005). 30% of the total fatalities (officially 87,350) had been victims of co-seismic landslides due to the Kashimir earthquake (Ms=7.6) (Havenith and Boureau 2010). In this chapter, we take the 2008 Wenchuan Earthquake (Ms=8.0) as example to discuss this issue.

2.1 The 2008 wenchuan earthquake

The earthquake had a magnitude of 7.9 Ms, occurred in Sichuan Province, China at 14:28 CST on 12 May 2008. The epicenter is located Yingxiu town (30.986°N, 103.364°E), Wenchuan County. The focal depth is about 12 km according to the report by the China Earthquake Administration (CEA).

The earthquake occurred along the Longmenshan fault (LMSF) zone at eastern margin of the Tibetan Plateau, adjacent to the Sichuan Basin as shown in Fig. 2 (Gorum et al., 2011). The fault belt is a series of faults striking in a northeast direction, on a North-South zone of high topographical and geophysical gradients between the Tibet Plateau on its western side and the Yangzi Platform on its eastern side. Seismic activities concentrated on its mid-fracture (known as Yingxiu-Beichuan fracture). Starting from Yingxiu, the rupture propagated unilaterally towards the northeast at an average speed of 3.1 kilometers per second, generating a 300-km and a 100-km long surface rupture along the Yingxiu-Berchuan and Pengguan faults, respectively (Huang et al., 2011a). The duration was as long as 120 seconds and the maximum displacement amounted to 9 meters.

Official figures, released by China News www.chinanews.com, on July 21, 2008 12:00 CST show that 69,197 are confirmed dead, 374,176 injured, and 18,222 listed as missing. The earthquake destroyed 5,362,500 and seriously damaged 21,426,600 houses, left about 4.8 million people homeless (Cui et al., 2009, Tang et al., 2011b). Approximately 15 million people lived in the affected area. It was the deadliest earthquake to hit China since the 1976 Tangshan earthquake, which killed at least 240,000 people.

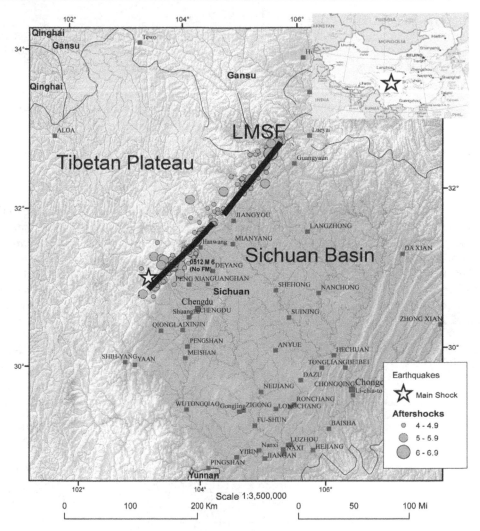

Fig. 2. The 2008 Wenchuan earthquake, LMSF and aftershocks. (modified from United States Geological Survey [USGS], 2008)

2.2 The landslides induced by the earthquake

The earthquake occurred in a mountainous region, where the geological and topographical features and climate conditions are very complex. The response to the ground shock was very strong. The recorded peak ground acceleration of local site reached to 2.0g (Huang et al., 2011b). Because of the complex terrain and climate conditions, the quake induced as many as 56,000 landslides (Dai et al., 2011). It is estimated that over one third of the total lost from the 2008 Wenchuan earthquake was caused by the earthquake induced landslides. Therefore, the secondary chain disaster induced by the 2008 Wenchuan earthquake is much more serious than the 1976 Tangshan earthquake.

2.3 Characteristics of the landslides

As many as 56,000 landslides have been identified by field investigations and using remote sensing technique with aerial photographs and satellite images. And the following distinctive characteristics can be summarized from these landslides:

1. Large scale

Many large scale landslides were induced by the 2008 Wenchuan earthquake. There are tens of landslides with a volume of 10^7 cubic meters (Wu et al., 2010), and 113 landslides with the area larger than 50,000 m^2 as shown in Table 1 (data from Xu et al., 2009a). The largest one is the Daguangbao landslide in Anxian County with an area of 7,273,719 m^2 and the volume of about $8.4 \times 10^8 m^3$ (Chigira et al., 2010).

2. The effect from the hanging and foot wall of the fault

It has been found that the majority of landslides are distributed in the range belonging to the hanging wall of the Yingxiu-Beichuan fault and Pengguan fault, northwest part of earthquake zone. All statistics seem to support that the landslide in hanging wall area is more active than in foot wall areas (Xu et al., 2009a; Yin et al., 2009a). For example, the distribution of large-scale landslides also shows the hanging / foot wall effect. It can be found from the Table 1 that 80 landslides occurred in the hanging wall, 70.8% of the total 113 large-scale landslides, and only 33 landslides occurred in the foot wall, 29.2% of the total number.

No.	Name	Place	Area /m²	Distance to fault /m	Wall location
1	Daguangbao	Anxian	7,273,719	4,800	Hanging wall
2	Wenjiagou	Mianzhu	2,945,520	3,900	Foot wall
3	Donghekou	Qiangchuan	1,283,627	300	Hanging wall
4	Zhengjiashan	Pingwu	1,014,987	2,400	Hanging wall
5	Shuimogou	Shifang	915,608	700	Hanging wall
6	Dawuji	Anxian	792,190	6,900	Hanging wall
7	Woqian	Qiangchuan	695,672	200	Hanging wall
8	Dashanshu	Mianzhu	693,687	6,900	Hanging wall
9	Hongshigou	Anxian	687,520	2,240	Hanging wall
10	bingkoushi	Pengzhou	575,556	12,600	Hanging wall
11	Tangjiashan	Beichuan	572,009	2,780	Hanging wall
12	Huatizigou	Pengzhou	541,193	4,980	Hanging wall
13	Wenjiaba	Pingwu	537,101	380	Hanging wall
14	Niujuangou	Wenchuan	527,700	300	Hanging wall
15	Haixingou	Mianzhu	517,573	8,888	Hanging wall
16	Ma'anshi	Pingwu	509,836	4,200	Hanging wall
17	Shibangou 1#	Qiangchuan	496,983	2,300	Hanging wall
18	Guershan	Beichuan	471,112	0	Hanging wall
19	Xiaojiashan	Mianzhu	465,899	2,900	Hanging wall
20	Xinkaidong	Pengzhou	449,685	6,800	Hanging wall
21	Boazangcun	Anxian	418,744	4,030	Hanging wall
22	Mianjiaoping	Beichuan	377,247	550	Foot wall
23	Weijiashan	Beichuan	358,021	2,120	Hanging wall
24	Liqigou	Jiangyou	355,113	10,000	Foot wall

No.	Name	Place	Area /m²	Distance to fault /m	Wall location
25	Caocaoping	Anxian	354,046	660	Hanging wall
26	Miepengzi 3#	Mianzhu	353,817	600	Hanging wall
27	Laoyinggou	Anxian	353,242	1,050	Hanging wall
28	Huoshigou	Anxian	322,155	1,400	Hanging wall
29	Zhangjiashan	Anxian	306,576	6,000	Foot wall
30	Macaotan	Mianzhu	305,989	2,700	Foot wall
31	Xiejiadianzi	Pengzhou	294,256	1,100	Hanging wall
32	Shibangou 2#	Qiangchuan	288,305	2,400	Hanging wall
33	Huishuituo	Pengzhou	270,980	4,200	Hanging wall
34	Dazhuping	Anxian	270,692	540	Hanging wall
35	Miepengzi 2#	Mianzhu	262,520	600	Hanging wall
36	Heshangqiao 3#	Dujaingyan	257,635	10,400	Hanging wall
37	Muguapingcun	Shifang	256,340	900	Foot wall
38	Miepengzi 1#	Mianzhu	255,296	600	Hanging wall
39	Dongxigou	Beichuan	246,020	2,200	Hanging wall
40	Yaozigou	Pingwu	242,553	800	Hanging wall
41	Baichaping	Dujaingyan	241,874	4,700	Hanging wall
42	Changping	Pengzhou	224,645	2,400	Hanging wall
43	Baodili	Qiangchuan	222,157	700	Hanging wall
44	Xiaomuling	Mianzhu	218,705	2,450	Hanging wall
45	Heshangqiao 1#	Dujaingyan	214,020	10,900	Hanging wall
46	Baishuling	Beichuan	208,968	4,350	Hanging wall
47	Dawan	Beichuan	203,959	2,150	Hanging wall
48	Baiguoshu	Beichuan	203,246	1,000	Hanging wall
49	Zengjiashan	Mianzhu	198,165	11,350	Foot wall
50	Zhangjiagou	Beichuan	196,299	640	Hanging wall
51	Zhaojiaqu	Qiangchuan	193,153	1,300	Hanging wall
52	Heitanzi	Anxian	182,452	8,900	Foot wall
53	Anleshan	Beichuan	180,809	1,140	Hanging wall
54	Yangshangou	Beichuan	177,361	1,300	Hanging wall
55	Xiaotianchi	Mianzhu	175,758	8,200	Foot wall
56	Yanyangcun	Beichuan	174,008	1,600	Foot wall
57	Shicouzi	Pingwu	169,540	0	Hanging wall
58	Chenjiaping	Anxian	169,368	1,050	Hanging wall
59	Wangyemiao	Dujaingyan	167,980	9,300	Hanging wall
60	Jiadanwan 1#	Dujaingyan	166,643	7,900	Hanging wall
61	Jinhelingkuang	Mianzhu	159,848	2,800	Foot wall
62	Fengyanzi	Beichuan	158,468	0	Foot wall
63	Changtan	Mianzhu	151,094	6,670	Foot wall
64	Weijiagou	Beichuan	150,818	450	Hanging wall
65	Xiaogangjian	Mianzhu	149,074	6,280	Foot wall
66	Baiyanshan	Qiangchuan	147,940	4,300	Hanging wall
67	Guoniucun	Beichuan	147,554	3,000	Hanging wall
68	Heshangqiao 2#	Dujaingyan	147,394	9,600	Hanging wall
69	Bazuofen	Anxian	146,272	11,000	Foot wall
70	Tiangengli	Qiangchuan	144,729	1,400	Foot wall

No.	Name	Place	Area /m²	Distance to fault /m	Wall location
71	Hongmagong	Qiangchuan	144,683	350	Foot wall
72	Baiguocun	Qiangchuan	139,800	300	Foot wall
73	Huangtuliang	Beichuan	135,084	550	Hanging wall
74	Qinglongcun	Qiangchuan	134,079	790	Foot wall
75	Pengjiashan	Beichuan	127,156	2,900	Hanging wall
76	Wangjiayan	Beichuan	125,381	400	Hanging wall
77	Yibadao	Mianzhu	125,059	9,600	Foot wall
78	Laohuzui	Wenchuan	125,039	2,700	Hanging wall
79	Beichuanzhongxuexinqu	Beichuan	124,365	300	Foot wall
80	Xiaomeizilin	Mianzhu	122,530	5,800	Foot wall
81	Xiangshuishi	Pengzhou	119,194	4,600	Hanging wall
82	Gaojiamo	Pingwu	115,301	1,600	Hanging wall
83	Jiadanwan 2#	Dujaingyan	114,905	9,300	Hanging wall
84	Dahuashu	Beichuan	113,111	0	Hanging wall
85	Wangjiabao	Beichuan	112,418	0	Hanging wall
86	Jiankangcun	Pingwu	111,106	340	Hanging wall
87	Xiaojiaqiao	Anxian	110,085	3,000	Foot wall
88	Lingtou	Qiangchuan	102,116	800	Hanging wall
89	Longwangou	Beichuan	99,821	650	Hanging wall
90	Zhangzhengbo	Qiangchuan	99,726	790	Foot wall
91	Nanyuecun	Dujaingyan	99,350	0	Hanging wall
92	Hongkouxiangxiajiaping	Dujaingyan	96,345	790	Hanging wall
93	Dujiayan	Qiangchuan	94,769	960	Foot wall
94	Madiping	Qiangchuan	94,633	2,600	Hanging wall
95	Maochongshan 1#	Pingwu	92,355	1,200	Hanging wall
96	Yandiaowo	Qiangchuan	92,128	340	Foot wall
97	Chuangzigou	Mianzhu	91,718	2,200	Foot wall
98	Xiaoxishan	Qiangchuan	90,298	1,000	Hanging wall
99	Xishanpo	Beichuan	83,663	1,140	Hanging wall
100	Hejiayuan	Qiangchuan	83,359	1,990	Foot wall
101	Zhaojiashan	Qiangchuan	82,329	1,000	Foot wall
102	Liushuping 1#	Qiangchuan	81,000	780	Hanging wall
103	Weiziping	Qiangchuan	74,661	470	Hanging wall
104	Gongziba	Qiangchuan	71,221	220	Hanging wall
105	Maerping	Qiangchuan	70,982	7,500	Hanging wall
106	Maochongshan 2#	Pingwu	70,252	1,200	Hanging wall
107	Muhongping	Qiangchuan	68,288	2,600	Foot wall
108	Machigai	Qiangchuan	66,602	500	Foot wall
109	Zixicun	Pingwu	57,820	2,400	Hanging wall
110	Liushuping 2#	Qiangchuan	54,810	1,000	Hanging wall
111	Dongjia	Qiangchuan	54,353	1,000	Foot wall
112	Majiawo	Qiangchuan	50,591	1,100	Hanging wall
113	Xiaowuji	Qiangchuan	50,122	2,100	Foot wall

Table 1. The large scale landslides with the area larger than 50,000 m² (data from Xu et al., 2009a).

3. The effect from the distance to the faults

Among the large scale landslides, the two farthest landslides from the fault are about 12.6km in the side of hanging wall and 11.35km in the foot wall. The majority of landslides (about 70%) occured in the region of 3km from the fault. Fig. 3a and 3b show the accumulative percentage of landslide distribution as a function of the distance to the fault in hanging and foot wall respectively. An exponential decay has been found for the number of landslides with the distance to the fault in both hanging and foot wall (Fig. 4).

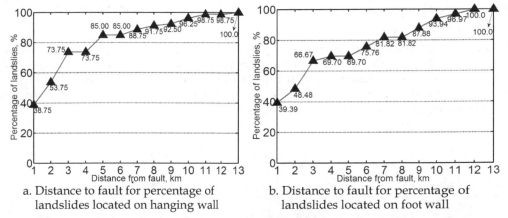

a. Distance to fault for percentage of landslides located on hanging wall

b. Distance to fault for percentage of landslides located on foot wall

Fig. 3. Relationship between percentage of landslides and large-scale distance to fault. a: Distance to fault for percentage of landslides located on hanging wall; b: Distance to fault for percentage of landslides located on foot wall.

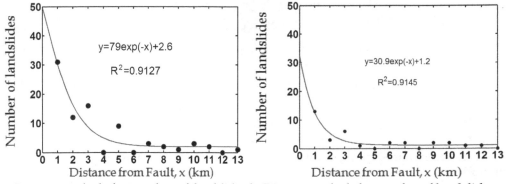

a. Distance to fault for number of landslides located on hanging wall

b. Distance to fault for number of landslides located on foot wall

Fig. 4. Relationship between number of landslides and distance to fault. a: Distance to fault for number of landslides located on hanging wall; b: Distance to fault for number of landslides located on foot wall.

4. Effect from the locking segment of the fault zone

The two largest scale landslides: Daguangbao landslide with the area of 7,273,719m² and Wenjiagou landslide with the area of 2,945,520m² are found locating at a distance of more than 3.9km from the fault from Table 1, although most of large-scale landslides, which many researchers have been studying on, are located in the region of less than 1km from the fault. For example, the Donghekou landslide (No.3 in Table 1) has a distance of 0.3km, the Woqian landslide (No.7 in Table 1) has a distance of 0.2km and Niujuangou landslide (No.14 in Table 1) has a distance of 0.3km from the fault. By examining the positions of the two landslides with the fault zone, it has be found that the two landslides are just located at the locking segment of the fault zone where high stress is believed to be concentrated and a lot of energy was absorbed by the locking of the rupture fault. Therefore, it should be notice that large scale landslides may occur at such kind of locking segment of the fault zone.

5. Direction effect

By examining the sliding directions of large-scale landslides along Hongshihe valley, it has been found that the directions parallel to or perpendicular to the fault are dominated as shown by the rose diagram in Fig. 5. It is implied that the landslides are controlled by the earthquake wave propagation and the fault movement. The slopes parallel to or perpendicular to the fault are easy collapsed.

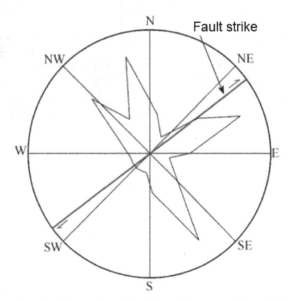

Fig. 5. Rose diagram showing the motion direction of large-scale landslides along the Hongshihe (after Xu et al., 2009a)

6. The long run-out characteristic

For rainfall induced landslides, the run-out distances are mostly less than 2 times of the slope height (2H). For example, 95% of the 19,035 landslides induced by rainfall are less than 50m based on the records from 1972 to 2008 in Japan. However, numerous rapid and long run-out landslides occurred during the 2008 Wenchuan Earthquake. Table 2 lists 21 long-run-out landslides with the horizontal distance larger than 500m. These landslides traveled over extraordinarily large distances with extremely high speeds and produced catastrophic results.

For instance, the Wangjiayan landslide (No. 76 in Table 1 and No. 21 in Table 2), occurred at the old town area of Beichuan city, had a run-out distance of 550m. It destroyed hundreds of buildings and resulted in more than 1600 fatalities (Yin et al., 2009a). The Daguangbao landslide (No. 1 in both Table 1 and 2) is another long run-out example. The affected area is estimated more than 7.2 km². Its run-out distance is estimated as 4,500m. The most complex in dynamic mechanism is the Donghekou landslide (No. 3 in Table 1 and No. 4 in Table 2) which has the run-out distance of over 2.4km. It blocked two rivers and formed two landslide lakes at Donghe village of Qingchuan County.

It has been found that the run-out distance is proportional to the area and volume of landslide. The regression formulas of $D/H=\lg S-3.12$ has been obtained with a coefficient of determination of $R^2=0.7681$, where D is the run-out distance, H is the slope height and S is the area of landslide. and $D/H=0.54\lg V-1.26$ with a coefficient of determination of $R^2=0.6290$, where V is the volume of landslide (see Fig. 6).

Since the mechanism of long run-out landslide is very important in landslide disaster mitigation, it will be discussed in Section 4.

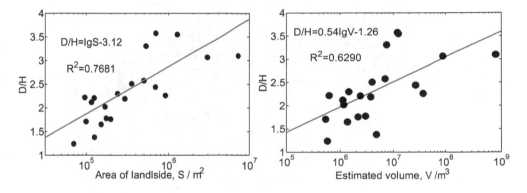

Fig. 6. The relationship between normalized run-out distance and (a) the area of landslide, S (m²), and (b) the volume of landslide, V (m³).

7. Large number of landslide dams

There are 34 landslide dams formed by the earthquake induced landslides. These landslide dams blocked the major large rivers. The water impounded by landslide dams created dam lakes.

The largest scale landslide dam was formed by the Tangjiashan landslide. It blocked the upper portion of the Jianjiang River at a location of about 5 km from Beichuan County Town. The dam crest extended approximately 600 m across and 800 m along the valley (Xu et al., 2009b). The maximum height of the dam is about 124 m. The maximum capacity of the landslide lake was 2.4×10^8m³, with the length of 20 km (Yin et al., 2009a).

Because of its rather loose nature and absence of controlled spillway, it is feared that the landslide dam may fail catastrophically and lead to downstream flooding with high casualties. Hundreds of thousands of residents in downstream Mianyang City were evacuated to the higher locations out of the town before a temporary drainage channel was digged in the dam by Chinese government.

Rank	No. in Table 1	Name	Place	Area/m²	Distance from fault/m	Estimated Volume/m³	Height H/m	Horizontal distance D/m	D/H	Wall location
1	1	Daguangbao	Anxian	7,273,719	4,800	7.5×10⁸※-8.4×10⁸◎	≈1450♦	4,500※	3.10	Hanging wall
2	2	Wenjiagou	Mianzhu	2,945,520	3,900	2.75×10⁷-1.5×10⁸※	1,360	4,170	3.07	Foot wall
3	14	Niujuangou	Wenchuan	527,700	300	7,500,000	827	2,638-3,102♦	3.31	Hanging wall
4	3	Donghekou	Qiangchuan	1,283,627	300	1.0×10⁷☆-1.5×10⁷	680	2,413	3.55	Hanging wall
5	24	Liqigou	Jiangyou	355,113	10,000	4,000,000	920	2,310	2.51	Foot wall
6	9	Hongshigou	Anxian	687,520	2,240	26,000,000▼	≈900♦	2,200▼	2.44	Hanging wall
7	5	Shuimogou	Shifang	915,608	700	36,000,000▼	≈930♦	2,100▼	2.26	Hanging wall
8	7	Woqian	Qiangchuan	695,672	200	12,000,000	570	2,043	3.58	Hanging wall
9	17	Shibangou 1#	Qiangchuan	496,983	2,300	7,000,000	710	1,681♦-1,829	2.58	Hanging wall
10	31	Xiejiadianzi	Pengzhou	294,256	1,100	3.5×10⁶-4.0×10⁶☆	740	1,500-1,750▼	2.19	Hanging wall
11	58	Chenjiaping	Anxian	169,368	1,050	1,163,446	680	1,372	2.02	Hanging wall
12	41	Baichaping	Dujiangyan	241,874	4,700	1,449,340	580	1,340	2.30	Hanging wall
13	63	Changtan	Mianzhu	151,094	6,670	1,367,868	800	1,320	1.65	Foot wall
14	49	Zengjiashan	Mianzhu	198,165	11,350	2,166,334	700	1,230	1.76	Foot wall
15	55	Xiaotianchi	Mianzhu	175,758	8,200	2,999,540	630	1,120	1.78	Foot wall
16	89	Longwangou	Beichuan	99,821	650	540,382	520	890	1.71	Hanging wall
17	82	Gaojiamo	Pingwu	115,301	1,600	1,126,823	340	722	2.12	Hanging wall
18	92	Hongkouxiangxiajiaping	Dujiangyan	96,345	790	624,296	300	666	2.22	Hanging wall
19	79	Beichuanzhongxuexinqu	Beichuan	124,365	300	2,400,000☆	≈300♦	664♦	2.21	Foot wall
20	106	Maochongshan 2#	Pingwu	70,252	1,200	581,719	490	610	1.24	Hanging wall
21	76	Wangjiayan	Beichuan	125,381	400	4,800,000☆	≈400♦	550☆	1.38	Hanging wall

Note: ※ data from Huang et al. (2011a); ▼ data from Wu et al. (2010); ♦ data from Qi et al. (2011); ◎ data from Chigira et al.(2010); and ☆ data from Yin et al.(2009a); ♦ Estimated indirectly from geological sections or description. ☆ notes the value is mean one if the number of data more than one.

Table 2. Some long run-out landslides triggered by the Wenchuan earthquake (arranged from the Horizontal distance) (modified from Xu et al., 2009a)

3. Susceptibility analysis of earthquake induced Landslides

Landslide susceptibility analysis (LSA) is necessary and important for land use planning and disaster mitigation. Many researchers have made great effort to identify the relationships of landslide characteristics such as distribution pattern, type, area coverage and volume with the triggering factors such as the magnitude, intensity and peak ground acceleration (PGA) of the earthquake, coseismic fault rupture (e.g. Lee et al., 2008; Rodriguez et al., 1999; Miles and Keefer, 2009; Keefer, 1984, 2000, 2002; Papadopoulos and Plessa, 2000). Some researchers have studied the relationships of landslide distribution with geo-environmental factors such as lithology, morphology, presence of secondary active or inactive faults (e.g. Chigira and Yagi, 2006; Jibson et al., 2000; Khazai and Sitar, 2003; Keefer, 2000; Yagi et al., 2009).

The Wenchuan earthquake induced landslides has been carried out by several researchers. For example, Huang et al. (2011a) studied the characteristics and failure mechanism of Daguangbao Landslide, the largest scale landslide, and suggested a classification system. Tang et al. (2011b) studied the effect of the quake on the landslides induced by the subsequent strong rainfall after earthquake by a case study in the Beichuan area. Qi et al. (2010) built a spatial database of landslides by using the remote sensing (RS) results which cover 11 counties seriously damaged by the earthquake. Yin et al (2009a, b) analyzed the landslide distribution, the mechanisms of some typical landslides, and evaluated the potential hazards of the landslide dams. Gorum et al. (2011) presented the preliminary results of an extensive study of the mapping the distribution of landslides by using a large set of optical high resolution satellite images. Yin et al. (2010) presented a quantitative result of the number and area of the landslides from Anxian to Beichuan. Dai et al. (2011) mapped over 56,000 landslides using aerial photographs and satellite images and characterized the spatial distribution of landslides by correlating landslide-point density and landslide-area density with the physical parameters that control the seismic stability of slopes.

In this chapter, we show some results from landslide susceptibility analysis carried out in Qingchuan County. Our analysis was based on slope units rather than the traditional grid units. At first, the relationship of landslide distribution with an individual causative factor is analyzed. And then, landslide susceptibility is analyzed by using artificial neural network (ANN) method. Finally, a landslide susceptibility map is made based on the ANN results.

3.1 Study area and data source

Qingchuan County is located at the north-western part of the earthquake zone as shown in Fig. 7. The landslides in the area of 3,271km^2 are studied.

3.2 Slope unit

Up to now, most of such studies were carried out based on the grid units. There is a problem in grid-based study that a grid may contain different slopes and a large slope may contain several grids with different slope grades. Despite the problem, the grid units were still used just because the slope units are difficult to be indentified for a wide range in the past.

Nowadays, it becomes possible and easy to indentify slope units by using GIS-based hydrologic analysis tool (David, 2002), which is based on the watershed divide and drainage lines. The slope unit size should be determined when the tool is used. We suggest that the appropriate slope unit size should match the average size of the landslide bodies in the study area.

A total of 55,899 slope units were indentified in Qingchuan County (Fig. 8). They will be used for landslide susceptible analysis in this study.

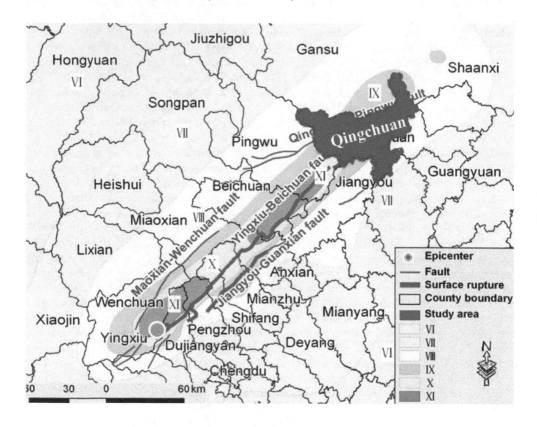

Fig. 7. Location of the study area

The basic data include a 1:100,000 geological map and a 10m grid digital elevation model (DEM) made from the available topographic map with 5m contour line interval. 885 landslides were identified from field investigations and RS results.

3.3 Relationship between landslide and individual causative factor
More than fifty factors can be considered as the landslide causative factors (Lin, 2003). By considering the data availability, analysis effectiveness, and independence of each factor (Lee et al., 2008), we selected the follow 8 factors: slope gradient, elevation, slope range, slope aspect, specific catchment area, lithology, distance to the fault and distance to the stream, to examine the causative factors contributing to the initiation of landslides.

Each causative factor was classified into several categories. The number of the slope units in each category is calculated and the percentage of the category among the whole slopes is given in Fig.9(b). For each category, the percentage of the failure slopes among the slopes in the same category is given as the landslide frequency in Fig. 9(a).

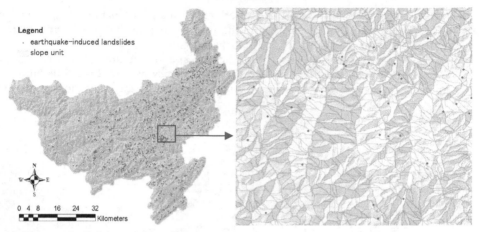

Fig. 8. Slope unit division of Qingchuan area

From the statistical analysis, the following results can be found.
1. More than 90% of the slopes have the slope gradient larger than 20°. The landslides occurred majorly in the slopes with gradients between 20° to 35°.
2. The landslides occurred majorly in the area with the elevations less than 1,200m.
3. The landslides occurred majorly in the slopes with slope ranges from 200 to 400m.
4. There is no clear relationship between landslides and specific catchment area.
5. The number of the landslides in the slopes in N direction is as twice as the slopes in the other directions.
6. The number of the landslides in the slopes with the distances to the fault less than 0.5km is as twice as the slopes in other categories.
7. The number of the landslides in the slopes with the distances to a stream less than 5km is as 3 times as the slopes in other categories.
8. There is no clear relationship between landslides and lithology.

3.4 Landslide susceptibility analysis using artificial neural network
The landslide susceptibility analysis is carried out by using artificial neural network (ANN) based on the above statistical analysis results.

ANN program is a "computational mechanism able to acquire, represent, and compute a mapping from one multivariate space of information to another, given a set of data representing that mapping, which is independent of statistical distribution of the data, can resolve the nonlinear problem and get high prediction accuracy for classification problem especially for large amount samples (Garrett, 1994). The applications of ANN to landslide susceptibility evaluation have been made by many researchers (e.g. Ermini,L., et al., 2005; S. Lee et al.,2006; Pradhan.B et al.,2010). Nefesilioglu et al. (2008) showed that ANN could give a more optimistic evaluation of landslide susceptibility than logistic regression analysis. Ermini et al. (2005) compared two neural architectures: probabilistic neural network and multi-layered perceptor, and obtained a better prediction result.

In this study, the neural network tool SPSS clementine is used since very few parameters are required. One group of the total slopes are randomly selected for training. 611 collapsed slopes of 885 landslides (70%) and 3300 of 55014 un-collapsed slopes (6%) are randomly selected for this group.

Two cases have been analyzed. Case 1 used the 8 factors mentioned as the statistical analysis and Case 2 used the 5 factors, 3 factors with the smaller weight values ware removed from the 8 factors. Also, different layers are used for the two cases.

The weights for each factor, experiments structures and the accuracy from the analysis results are shown in Table 3. It can be seen that the accuracy of Case 2 is a little bit higher than Case 1. Therefore, the ANN model from Case 2 is used for the landslide susceptibility classification.

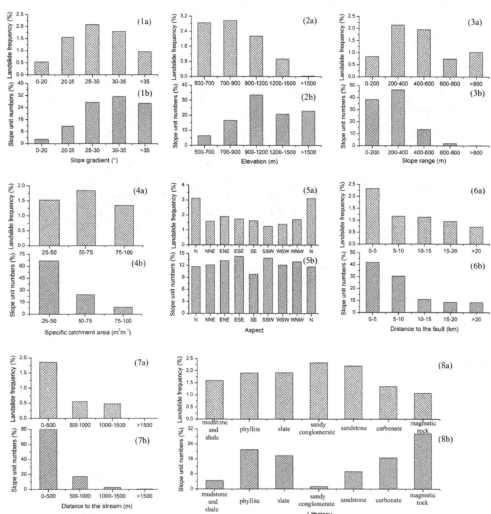

Fig. 9. Landslide frequency and slope unit numbers (%) for each category of causative factor

The output landslide susceptibility indices (LSI) were converted to GIS grid data in three susceptible levels as shown in Table 4. There are 4145 slopes identified as high susceptible level (dangerous slopes), 48,373 slopes identified as low susceptible level (stable slopes), 3,382 slopes identified as medium susceptible level (gray zone).

Causative factors	Case 1	Case 2
Slope gradient	0.367	0.446
Elevation	0.253	0.327
Slope range	0.193	0.163
Aspect	0.038	×
Specific catchment area	0.037	×
Distance to the fault	0.054	0.016
Distance to the stream	0.035	0.048
Lithology	0.023	×
Accuracy	95.05%	96.39%
ANN Structure	8*16*1	5*3*5*1

Table 3. Weight of each factor in 2 cases

By comparing with real landslides, it can be found that 877 of 885 landslides and 3268 stable slopes are identified as high susceptible level, which means 99% of landslides can be predicated by the model but 78.8% of predictions would be false alarm.

On the other hand, 1 of 885 landslides and 48373 stable slopes are identified as low susceptible level, which means 99.4% of predications are correct and less than 0.2% landslides would not be alarmed.

LSI	Susceptible level	Practical		Analysis result	
		Collapsed	Stable	Slope number	(%)
0.0-0.01	Low	1	48372	48373	96.53
0.01-0.1	Medium	7	3375	3382	6.05
0.1-0.9966	High	877	3267	4144	7.42
	Total	885	55014	55899	100.00

Table 4. Characteristics of the three susceptibility zones

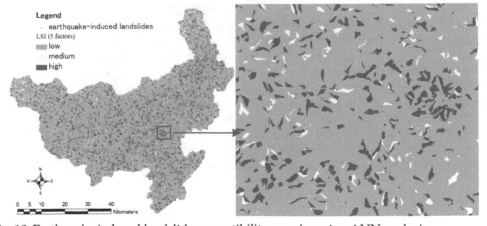

Fig. 10. Earthquake-induced landslide susceptibility map by using ANN analysis.

The landslide susceptibility map is made from the ANN results (Fig. 10). The high susceptible zone occupies 7.42%, the low susceptible zone 86.53% of the whole area. In addition, 6.05% of the area is gray zone.

3.5 Conclusions

Landslide susceptible analysis has been carried out in Qingchuan County. 55,899 slope units have been extracted and used for the analysis. The relationship between landslide distribution and the individual causative factor has been investigated by statistical analysis. The clear relationship can be identified for slope gradient, elevation, slope rang, the distances to the fault and the distances to a stream. The ANN analysis also showed the same results, that is, slope gradient, elevation, slope range, distance to the fault and distance to a stream have relatively larger weight. By removing the other three factors with smaller weights, the ANN analysis accuracy got improved. By comparing landslide occurrence locations with susceptibility zones, it has been shown that 99% of landslides can be predicated by the obtained ANN model, but 78.8% of predictions would be false. On the other hand, 99.4% of stable predications are correct and less than 0.2% landslides would not be alarmed. In addition, the gray zone occupies 6% of the whole area. Therefore, the landslide susceptibility classification presented in this study is acceptable.

4. Analysis of Long run-out mechanism based on trampoline effect

The estimation of the movement behaviour of a potential landslide is very important in order to mitigate the landslide disaster. Especially, the run-out distance is one of the major parameters in landslide risk assessment and preventive measure design. Long run-out is one of the major characteristics of earthquake induced landslides. However its mechanism has not been understood very well.

Many researchers have made great effort to understand how and why large falling masses of rock can move unusually long run-out distance. Researchers have repeatedly revisited the problem using a wide variety of approaches. These efforts have yielded no less than 20 mechanical models for explaining long run-out in high-volume rapid landslides. Shaller and Shaller (1996) made a detail summary of the existed models and divided these models into four categories (1) bulk fluidization and flow of landslide debris; (2) special forms of lubrication along the base of the slides; (3) mass-loss mechanisms coupled with normal frictional sliding; and (4) individual-case mechanisms.

Actually, most of the existed models are helpful in the estimation of the run-out distance. However, very less of them considered the earthquake dynamical behaviour. For this reason, in this study, we take into account the so-called trampoline effect of earthquake on landslides and propose a multiplex acceleration model (MAM) to explain the long run-out mechanism. Since the MAM model can be easily incorporated into numerical methods, it can be applied to simulate the long run-out landslide very well.

4.1 Multiplex acceleration model

For an earthquake induced landslide, the following effects on the movements of the falling stones from the landslide can be considered: (1) a falling stone can obtain kinetic energy from the colliding with the vibrating slope during earthquake; (2) the force of friction between a falling stone and the slope can decrease since the normal force varies with the contact condition during earthquake; (3) The flying and rotation movement of a falling stone may occur much easily in earthquake induced landslides.

In order to consider these effects, we divide a period of wave is divided into two phases: P-phase and N-phase as shown in Fig. 11. The P-phase is defined as the period when the slope is moving in the outer normal direction of the slope surface. The slope is pushing the falling stones on its surface and lets them obtain kinetic energy in the P-phase. The N-phase is defined as the period when the slope is moving in the inner normal direction of the slope surface. Since the normal force will decrease when the slope surface moves apart from the falling stones, the force of friction will get decreased in the P-phase.

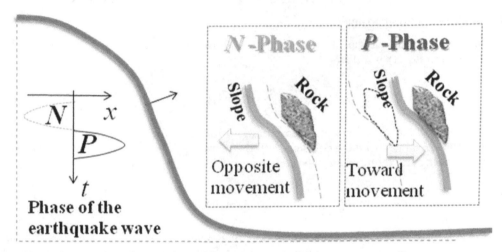

Fig. 11. P-phase and N-phase definition in MAM

By the repeated exchange of two phases during an earthquake, the falling stone get multiplex accelerated. The MAM model can be seen more clearly by apparent friction angle analysis.

Supposing that a stone with mass m moves from position A to position B during a landslide without earthquake (see Case 1 in Fig. 12), the potential energy decreases by mgh. Based on the energy conservation law, it is easy to obtain the following equation for a falling stone movement in the case without earthquake.

$$mgh - \sum_{i=1}^{n} l_i mg k_i \tan \varphi_{si} \cos \theta_i = 0 \tag{1}$$

The first term here is for potential energy and the second term is for the work of friction force between the slope and the falling stone, where the sliding movement is considered and the whole curve path is divided into finite linear segments. And m = mass, g = gravity acceleration, h = the falling height, l = the segment length, θ is the segment slope angle, φ is the friction angle, k is the coefficient of conveying from static to dynamic friction and i is the index of segment.

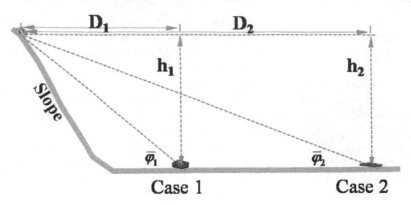

Fig. 12. Apparent friction angle

The apparent friction angle, usually used for the discussion of run-out distance, can be obtained from Eq. (1) as follows

$$\tan \overline{\varphi_1} = \frac{h_1}{D_1} = \sum_{i=1}^{n} w_i k_i \tan \varphi_{si} \tag{2}$$

When we consider the effects of slope vibration due to earthquake, the kinetic energy of falling stone obtained from the collision with the vibrating slope and the movement patterns (sliding, rolling and flying) should be considered. Thus, Eq. (1) becomes

$$mgh + \sum_{j=1}^{m} \frac{1}{2} m v_{ej}^2 - \sum_{i=1}^{n} l_i mg k_i^* \tan \varphi_{si} \cos \theta_i = 0 \tag{3}$$

The second term here is for the kinetic energy of a falling stone obtained from the collision with the vibrating slope and v_{ej} is the velocity obtained in jth P-phase and can be expressed as follows

$$v_{ej} = VTR \int_{t_j}^{t_j + \Delta t} f(t) dt \tag{4}$$

$f(t)$ is the acceleration of slope vibration due to earthquake, VTR is called the velocity transmission ratio due to collision.

The apparent friction angle for the case 2 in Fig. 12 can be obtained from Eq. (3) as follows

$$\tan \overline{\varphi_2} = \frac{h_2}{D_2} = \sum_{i=1}^{n} w_i k_i^* \tan \varphi_{si} - \sum_{j=1}^{m} \frac{v_{ej}^2}{2g D_2} \tag{5}$$

Comparing Eq.(5) with Eq.(2), it can be seen clearly that the mechanism of long run-out distance is as follows.

1. The kinetic energy of a falling stone obtained from the collision with the vibrating slope may result in long run-out distance from the second term of Eq. (5).
2. The coefficient of conveying from static to dynamic friction k^* in Eq. (5) can be smaller than k in Eq. (3) because of the N-phase effect, air cushion effect, movement pattern.

4.2 Colliding effect

In P-phase, a falling stone can obtain kinetic energy from the colliding with the vibrating slope. According to elastic collision theory, when two objects with different masses collide with each other, the object with smaller mass could obtain larger velocity. Since the mass of a slope is much larger than the mass of a falling stone, the velocity of the falling stone can be much larger than the vibrating velocity of the slope. That is to say the VTR in Eq. (4) can be larger than 1.0.

The VTR can be examined by the simple model shown in Fig. 13(a) and (b). The masses of the two blocks are m_1 and m_2 respectively. Before the colliding, the block 1 has initial velocity V_{10} toward block 2 which is standstill, i.e. $V_{20}=0$. The friction between blocks and the base is negligible. After the colliding, the velocity of block 1 becomes V_{11} while block 2 obtains a velocity V_{21}.

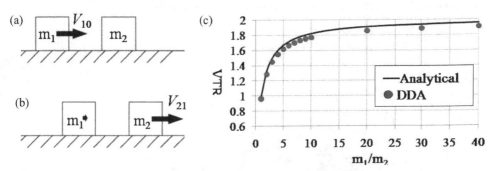

Fig. 13. The colliding model and VTR with mass ratio. (a) before the collision, (b) after the collision, (c) VTR obtained by DDA comparing with analytical solution.

According to the principles of the conservation of both energy and momentum, we have the following equations.

$$m_1 V_{10}^2 + m_2 V_{20}^2 = m_1 V_{11}^2 + m_2 V_{21}^2 \qquad (6)$$

$$m_1 V_{10} + m_2 V_{20} = m_1 V_{11} + m_2 V_{21} \qquad (7)$$

By solving Eqs. (6) and (7), we can obtain the VTR for the case of $V_{20}=0$ as follows

$$VTR = 2 \cdot \frac{m_1}{m_1 + m_2} \qquad (8)$$

It can be seen from the analytical solution Eq. (8) that if m_1 is much larger than m_2, VTR is to approach to 2.0. Therefore, since the mass of a slope is far larger than the falling stone, the velocity of the falling stone obtained from the slope vibration will be two times of that of the slope vibration velocity during earthquake.

The results of VTR given in Eq. (8) have been verified by DDA simulation. The model shown in Fig. 13 is used in DDA simulations. The block one with mass of m_1 has the initial velocity of 10 m/s and the block two with mass of m_2 is at a standstill. After the block one impacted the block two, the velocities of both blocks changed. The block two obtained the velocity from block one. The VTR is calculated from the ratio of V_{21} to V_{10}.

The results obtained from DDA simulations by changing m_1 are shown in Fig. 13(c), together with the theoretical analytical values. The line is calculated from the analytical solution Eq.(8) and the dots are obtained from DDA simulations.

It can be seen that the *VTR* obtained from DDA is in quite good agreement with the analytical solution. However, by close examination, it can be found that the *VTR* values from DDA are little smaller than the analytical values when the mass ratio of m_1 to m_2 is larger than 4.0. This is because elastic strains of the two blocks are led to energy transformed into potential energy of deformation by the collision in DDA simulation while no strain is considered in analytical solution.

Furthermore, when the block 2 has an initial velocity toward block 1, the *VTR* could become the larger and larger. Fig. 14 shows the results from DDA simulation. This may happen when a stone fall down to the slope in a *P*-phase, it will get larger rebounding velocity. That means, a trampoline effect can be produced by strong earthquake.

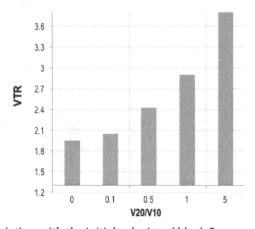

Fig. 14. The VTR variation with the initial velocity of block 2

4.3 Model tests by shaking table
Model tests using shaking table were carried out in order to investigate the effects of earthquake on the movement of debris. The model slope has the height of 180cm and the slope angle can be adjusted from 30° to 35° as shown in Fig. 15.

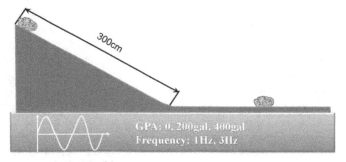

Fig. 15. Model tests by shaking table

The movements of 4 kinds of stones with different shapes have been investigated under the earthquake conditions of 0, 200gal and 400gal sine wave of 3Hz. More than 10 times of repeated experiments have been carried out for each case. The following results have been obtained.

1. The movement distance for the case of a 400gal earthquake is 3.4 times longer than the case of no earthquake for the No.4 stone. Therefore, the effect of earthquake on the movement distance is very large.
2. The movement distance for the case of a 400gal earthquake is longer than the case of a 200gal earthquake. So it seems that the movement distance is proportional to the earthquake magnitude.
3. The shape of the falling stone has effect on the movement distance. The movement distance of the No. 5 stone is much smaller than that of No. 4. This is because that the earthquake can change the movement pattern and cause the rotation motion. It can be seen that the No. 5 stone has very sharp edges and vertices, which may stop its rotation movement (Fig. 16).

It should be noticed that it is difficult to distinguish the velocity obtained from P-phase because the model slope is too small and there are very few P-phase during the movements.

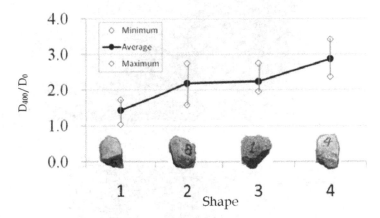

Fig. 16. Difference from shape of falling stone

4.4 Numerical simulation of landslide by using DDA

Simulation of landslide by using numerical methods is an effective way in order to overcome the dimension limit of the model test by using shaking table. In this study, Discontinue Deformation Analysis (DDA), developed by Shi and Goodman (Shi et al., 1984), is used since it is applicable to simulating the rigid body movements and large deformations of a rock block system under general loading and boundary conditions. Several extensions of the original DDA have been made in this study so that earthquake wave can be taken into the simulation for different ways.

Before simulating a real landslide, the applicability of the extended DDA has been verified by various simple models with theoretic solutions. For example, a simple model shown in Fig. 17 is calculated by both the theoretical solution and DDA simulation.

The theoretical solution of movement distance can be calculated by the following formula:

$$S_0 = \frac{1}{2}at^2 \tag{9}$$

where

$$\alpha = g[\sin\theta - (k\tan\varphi_0)\cos\theta] \tag{10}$$

The results of the movement distance are in good agreement with each other as shown in Fig. 18.

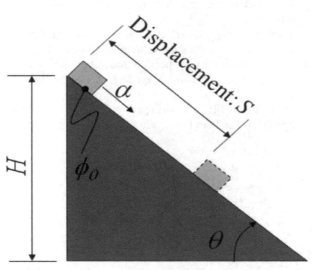

Fig. 17. The DDA model

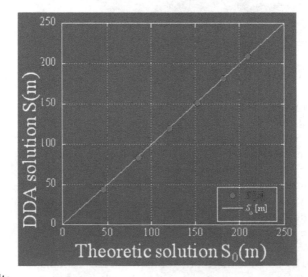

Fig. 18. The results

We applied the extended DDA to simulate the Dongheko landslide in Qingchuan prefecture. The vertical section shown in Fig. 19 is taken along the red line in Fig. 19. The DDA software and the model are shown in Fig. 20. The parameters for both the material and DDA program are also shown in Fig. 20.

Since the real earthquake curves are not available, a sine wave is used. The movements of debris at different times obtained from DDA simulation are shown in Fig. 21.

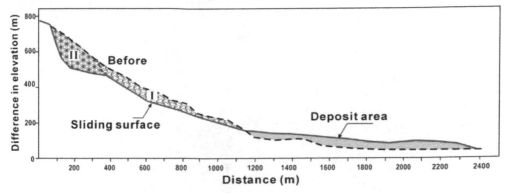

Fig. 19. Vertical section of the Donghekou landslide

It has been shown that an 800gal sine wave can cause long distance movements of debris like real one. The rotation and flying movements are the major reasons for long-distance movement, which can be easily observed in DDA simulations.

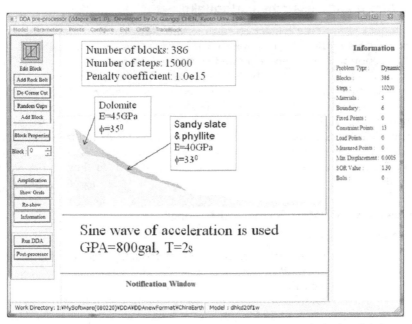

Fig. 20. DDA software developed by Chen and the model of Dongheke landslide

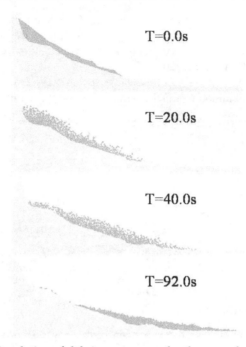

T=0.0s

T=20.0s

T=40.0s

T=92.0s

Fig. 21. The numeric simulation of debris movements by the extended DDA

5. Debris flow arising from the earthquake

Strong earthquakes not only trigger co-seismic landslides but also they affect subsequent rainfall-induced debris flows over a long term because these co-seismic landslides greatly increased the amount of sediment material for potential debris flows (Lin et al.,2006; Tang et al.,2009; Khattak et al.,2010). After the 2008 Wenchuan Earthquake, the earthquake affected areas experienced two rainy seasons till 2010, and a large number of debris flows occurred, which claimed as many as 450 fatalities. It makes the restoration and reconstruction much more difficult (Xie et al.,2008).

In this section, at first, the characteristics of debris flows in the earthquake affected areas are summarized. And then, an approach of simulating debris flow is proposed for disaster mitigation. Finally, a large scale debris flow is simulated so as to show the effectiveness of the proposed.

5.1 Characteristics of debris flow after the earthquake

1. Clear relation to the earthquake

A large number of debris flows occurred in the earthquake affected areas. For example, there are 46 debris flows found in Beichuan area. The distribution of these debris flows is shown in Fig. 22 in which the red line indicates the main fault of the quake, blue lines indicate rivers and the numbers indicate the locations of the debris flows. It can be seen that these debris flows are distributed along the rivers on the two sides of the earthquake fault. Therefore, the debris flows are highly related to the earthquake.

2. Large surge peak discharge and huge volume

Since the material sources of debris flow got much richer after earthquake, it is easy to form large scale debris flow. For example, the surge peak discharge reached 260 m^3/s in the debris flow occurred in Beichuan town on Sept. 24, 2008. The volume was too large to a basin with the area of 1.54km^2. The cover of debris is so thick that it buried the fourth floor of some buildings. Another example is Sanyanyu debris flow. The volume of the debris reached 144.20 million m^3. The debris flow carried many huge stones and destroyed houses and bridges (Tang et al., 2009).

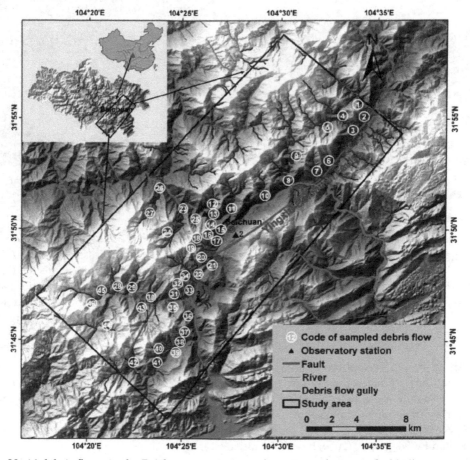

Fig. 22. 46 debris flows in the Beichuan area(Picture from Tangchuan et cl., 2010)

Many preventive structures designed based on the standard of conventional debris flow were also destroyed by the large scale debris flows after the earthquakes. For example, 19 check dams were destroyed by the Wenjia debris flow occurred in Mianzhu Qingping town area on Aug. 13, 2010. The Fig. 23(a) shows one of the check dam destroyed by the debris flow. The extreme large scale and destructive impact of the debris flow seems beyond imagination.

3. The low critical precipitation for triggering debris flow

The critical precipitation for triggering debris flow got decreased obviously after the earthquake. For example, the critical precipitation of 37mm became lower after the earthquake in Zhouqu County areas. A 22mm rainfall could trigger a debris flow during the past 3 years. According to preliminary analysis by Tang et al. (2009), the critical cumulative precipitation has been reduced about 14.8%-22.1%, the critical rainfall intensity per hour about 25.4 % ~31.6% in Beichuan County area.

Fig. 23. (a): The destroyed check dam in Qingping debris flow; (b): Ming river blocked at Yingxiu town by Hongchungou debris flow (photographs from Tang chuan)

4. River blocking

The disasters chain induced by the earthquake is very significant. The earthquake induced landslides caused debris flows which blocked rivers, and flooding disasters occurred. For example, Jianjiang River was blocked at 3 locations and half blocked at 8 locations by debris flows during the rainstorm on Sept. 24, 2008. Mianyuan River was blocked at 2 locations and half blocked at 11 locations by debris flows occurred on Aug. 13, 2010. Ming River was blocked at 1 location and half blocked at 5 locations by debris flows occurred on Aug. 14, 2010 (see Fig. 23 (b)).

5.2 An approach of simulating debris flow

Many studies on debris flow focused on estimation of an alluvial fan for predicting debris flow inundation areas (Glade,2005;Berti and Simoni,2007). They can be divided into three categories: dynamic models, volume-based models and topographic based models. For the 2008 Wenchuan earthquake, an empirical formula of estimating alluvial fan has been presented by Tang et al. (2010) based on statistical analysis.

In this study, we propose an approach for numerical simulation of debris flow to analyze or predict the peak discharge and the volume of a debris flow and its inundation area. The approach consists of the following 5 procedures.

1. Identify the earthquake induced landslides

The earthquake induced landslides can be identified by RS technique using aerial photos and satellite imagines. The locations and the shapes of all the debris deposits in a drainage area should be obtained from this procedure.

2. Make field investigations.

The thicknesses of all the debris deposits and the geological and geotechnical behaviors should be investigated in this procedure.

3. Generate the grids using GIS.

Grids are required for solving equations with finite different method. A DEM map can be converted to a raster image using GIS for the drainage area. The grids can be obtained by saving the raster data.

4. Solve the equations.

The debris and water mixture is assumed to be a uniform continuous, incompressible, unsteady Newtonian fluid. The following Navier-Stokes equations are used for debris flow governing equations:

$$\nabla u = 0$$
$$\rho \frac{\partial u}{\partial t} + \rho u \cdot \nabla u = \rho g - \nabla p + \mu \nabla^2 u \tag{11}$$

where $u = (u,v,w)$ is velocity; ρ is the mass density; p is the pressure; μ is dynamic viscosity; $g = (0,0,g)$, g is the gravitational constant and t is time.

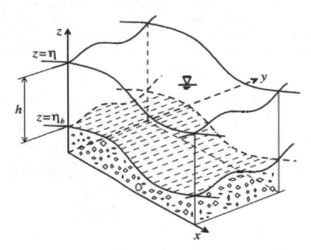

Fig. 24. Definition of coordinate system for 2D governing equations

The so-called depth-averaged model as shown in Fig. 24 is used. And then the following equations are used instead of Eq. (11). They are solved by finite difference method (FDM).

$$\frac{\partial h}{\partial t} + \frac{\partial M}{\partial x} + \frac{\partial N}{\partial y} = 0 \tag{12}$$

$$\frac{\partial M}{\partial t} + \alpha \frac{\partial (MU)}{\partial x} + \alpha \frac{\partial (MV)}{\partial y} = -\frac{\partial H}{\partial x} gh + v\beta(\frac{\partial^2 M}{\partial x^2} + \frac{\partial^2 M}{\partial y^2}) - gh \cos\theta_x \tan\xi \tag{13}$$

$$\frac{\partial N}{\partial t} + \alpha \frac{\partial (NU)}{\partial x} + \alpha \frac{\partial (NV)}{\partial y} = -\frac{\partial H}{\partial y} gh + v\beta (\frac{\partial^2 N}{\partial x^2} + \frac{\partial^2 M}{\partial y^2}) - gh \cos\theta_y \tan\xi \qquad (14)$$

Where $M = Uh$ and $N = Vh$ are the $x-$ and $y-$ components of the flow flux; U and V are the $x-$ and $y-$ components of the depth-average velocity; H is the height of the free surface; h is the flow depth; θ_x and θ_y are the angle of inclination at the bed along the x and y directions, respectively; α and β are the momentum correction factors; $v = \mu / \rho_d$ is kinematic viscosity, ρ_d is the equivalent density of the debris mixture, and $\rho_d = \rho_s v_s + \rho_w v_w$, ρ_s and ρ_w are the densities of solid grains and water, v_s and v_w are the volumetric concentrations of solids particles and water in the mixture; and $\tan\xi$ is the dynamic friction coefficient.

5. Visualize the results

The results from FDM are converted to GIS layers. The maps of maximum surge peak discharge, velocity distribution and in inundation area can be made by GIS. Also the animation of debris flow can be easily made.

5.3 Numerical simulation of the Hongchungou debris flow

The proposed approach has been used to simulate the Hongchungou debris flow occurred in Hongchungou drainage area on August 14, 2010. The materials carried by the debris flow blocked Ming River just at a little upper side about 200m from Yingxiu town, the epicenter of the 2008 Wenchuan Earthquake (Fig. 25), . The road along the river became the new temporary river channel and water flowed into Yingxiu town. As the result, serious flood occurred in the newly reconstructed town. The disaster claimed 13 lives and 59 missing persons.

The distribution of the earthquake induced landslides has been identified by using RS with the aerial photographs taken by The Ministry of Land and Resources of China. The aerial photograph 0.3m resolution of Hongchungou area is shown in Fig. 26(a).

In this study, the object-based analysis (OBA) is adapted for imagine analysis unlike traditional spectral information based image analysis method since there exists the so-called 'salt and pepper' appearance in the output of the latter (Tapas R. Martha et al., 2010). The analysis includes the following steps.

1.   The aerial photos are ortho-rectified based on the 20m DEM obtained from the China Geology Survey Bureau, in order to remove the distorting effects of tilt and terrain relief.

2.   The image is divided into objects based on homogeneity of pixel values through edge-based segmentation algorithm (Kerle et al., 2009), since it is very fast and only one parameter is needed for scale level. Scale level 30 and merge level 93 are used for image segmentation.

3.   NDVI index is used to separate vegetation from other objects. Spatial, spectral and texture attributes are separately computed for each object. Then, various land covers are classifies based on user-defined training data, selected by combing with 3 dimensional image in order to improve the interpretation refinement;

4.   The non-landslide objects are eliminated by the assumption that landslide will not occur for the slope gradient less than 5°.

Fig. 25. The blocked Ming River and new Yingxiu town in flood by Hongchungou debris flow.

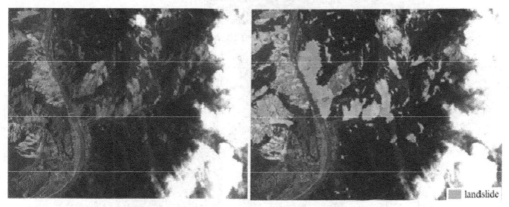

Fig. 26. (a): Aerial photography of Hongchun valley; (b): identified landslides in Hongchun valley

The obtained landslides are shown in Fig. 26(b). It can be seen that all the landslides were accurately recognized. There is not so-called 'salt and pepper' appearance.

Combining with field investigation, we finally selected four landslides: H1, H2, H3, H4 as the main loose source material of the debris flow (Fig. 27). The area of each landslide is 7,688 m² for H1, 5,137 m² for H2, 2,002 m² for H3, 4 4,567 m² for H4.

Since only a 20m DEM map is available for generating grids, the 4,000mX3,400m area is divided into 200X170 grids by using GIS. The rheological parameters are assumed constant (Tang et al., 2011a), and they are: ρ_d =2050kg/m³, α =1.25, β =1.0, μ =0.11, g = 9.8m/s², $\tan\xi$ =0.6.

The results from FDM are converted to GIS layers for visualizing. The movements of the debris flow are illustrated in Fig. 28(a)-(d) for different time. It can be seen that the river is blocked in Fig. 28(d). The distribution of the maximum depth of the whole flow is shown in Fig. 28(e). According to the simulation results, the debris flow takes 150s to travel about 3,300 m along the valley with an average flow velocity of about 22m/s.

Comparing the simulated results with the actual event, we found that they are in good agreement with each other. Therefore, the proposed approach has been shown applicable and useful for predicting the movement of potential debris flow arising from earthquake.

Fig. 27. Identified landslides as loose material of the debris flow

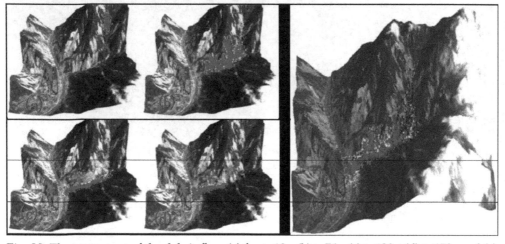

Fig. 28. The movement of the debris flow (a) for t=10s, (b) t=76s, (c) t=120s, (d) t=150s and (e) for the distribution of the maximum depth of Hongchungou debris flow.

6. Conclusions

A strong earthquake can induce a chain of disasters. The disaster chain from the 2008 Wenchuan earthquake has been discussed.

The characteristics of the earthquake induced landslides have been summarized as follows.
1.   A large number of landslides (56,000) were induced by the earthquake;
2.   Large quantities of large-scale landslides (113) occurred and are listed in this chapter.

3. The landslides in the hanging wall are more than the footing wall and about 70% landslides are located in the region of 3km from the fault.
4. Large-scale landslide can occur at the locking segment of the rupture fault.
5. There is a clear relationship between the sliding direction of landslides and the fault strike.
6. Large quantities of long run-out landslides occurred and are listed in this chapter. Regression formulas of run-out distance have been obtained based on the areas and volumes of landslides.
7. A large number of landslide dams (34) were formed.

The susceptibility analysis of the earthquake induced landslides has been carried out by both statistical analysis and ANN analysis based on slope units rather than the traditional grids. The relationship of landslide distribution with individual causative factor has been investigated. It has been found that slope gradient, elevation, slope range, the distances to the fault, the distances to a stream have contributed to landslides while specific catchment area, slope aspect and lithology have no clear relationship. A susceptibility map has been made for analysis of the earthquake induced landslides in Qingchuan County.

A Multiplex Acceleration Model has been proposed for analysis of the long run-out mechanism based on trampoline effect. Table model tests and DDA simulation were carried out. It has been shown that the proposed Multiplex Acceleration Model is reasonable and applicable.

The earthquake induced landslides can easily form debris flows after the earthquake. The characteristics of the debris flows arising from the 2008 Wenchuan earthquake have been summarized as follows.

1. There is a clear relation to the earthquake according to the distribution of the debris flows along the earthquake fault.
2. Most of debris flows have large surge peak discharges and huge volumes.
3. The critical precipitation for triggering debris flow became lower.
4. Many rivers were blocked by the debris flows and serious damages have been caused by the debris dams.

An approach of simulating debris flow has been proposed. The earthquake induced landslides are identified by RS with object-based analysis method, which can overcome the problem of the so-called 'salt and pepper' appearance existed in the traditional spectral information based image analysis method. A practical simulation has been carried out and the proposed approach has been shown effective and useful for estimating the movement behaviours of a potential debris flow arising from a strong earthquake.

7. Acknowledgment

The presented research work and the preparation of this paper have received financial support from the Global Environment Research Found of Japan (S-8), Grants-in-Aid for Scientific Research (Scientific Research (B), 22310113, G. Chen) from JSPS (Japan Society for the Promotion of Science). These financial supports are gratefully acknowledged.

8. References

Berti, M., Simoni,A. (2007). Prediction of debrisflow inundation areas using empirical mobility relationships.*Geomorphology,* Vol. 90,(Octorber 2007), pp. 144-161

Changa, KJ., Taboada, A., Linb, ML. & Chenc, RF. (2005). Analysis of landsliding by earthquake shaking using a block-on-slope thermo-mechanical model: Example of Jiufengershan landslide, central Taiwan. *Engineering Geology*, Vol. 80, (May 2000),pp. 151–163, ISSN 0013-7952

Chigira, M. & Yagi, H. (2006). Geological and geomorphological characteristics of landslides triggered by the 2004 Mid Niigta prefecture earthquake in Japan. *Engineering Geology*. Vol. 82, No. 4, (February 2006), pp. 202–221, ISSN 0013-7952

Cui, P., Zhu, Y., Han, Y., Chen, X. & Zhuang, J. (2009). The 12 May Wenchuan earthquake-induced landslide lakes: distribution and preliminary risk evaluation. *Landslides*, Vol. 6. No. 3, (June 2009), pp. 209-223, ISSN 1612-510X

Dai, F., Xu, C., Yao, X., Xu, L., Tu, X. & Gong, Q. (2011). Spatial distribution of landslides triggered by the 2008 Ms 8.0 Wenchuan earthquake, China. *Journal of Asian Earth Sciences*, Vol. 40, No. 4, (March 2011), pp. 883–895, ISSN 1367-9120

Dai, F.C., Lee, C.F. (2002). Landslide characteristics and slope instability modeling using GIS Lantau Island, Hong Kong. *Geomorphology* Vol. 42, (January 2002), pp. 213 – 238

David, R. M. (2002). *Arc Hydro: GIS for Water Resources*, ESRI Press, Redlands, CA,US.

Ermini, L., Catani, F., Casagli, N. (2005). Artificial neural networks applied to landslide susceptibility assessment. *Geomorphology* 66 (1 – 4), pp. 327 – 343

Garrett, J. (1994) Where and why artificial neural networks are applicable in civil engineering, *Journal of Computing Civil Engineering* , Vol.8, No.2, pp. 129–130

Glade,T.,2005.Linking debris-flow hazard assessments with geomorphology. *Geomorphology*, Vol. 66, pp. 189-213

Gorum, T., Fan, X., Westen, C., Huang, R., Xu, Q., Tang, C. & Wang, G. (2011). Distribution pattern of earthquake-induced landslides triggered by the 12 May 2008 Wenchuan earthquake. *Geomorphology*, doi:10.1016/j.geomorph.2010.12.030, ISSN 0169-555X

Huang, R., Pei, X., Fan, X., Zhang, W., Li, S. & Li, B. (2011a). The characterisitics and failure mechanism of the largest landslide triggered by the Wenchuan earthquake, May 12, 2008, China. *Landslides*, (June 2011), doi:10.1007/s10346-011-0276-6, ISSN 1612-510X

Huang, R., Xu, Q. & Huo, J. (2011b). Mechanism and geo-mechanics models of landslides triggered by 5.12 Wenchuan earthquake. *Journal of Mountain Science*, Vol. 8, No. 2, (April 2011), pp. 200-210, ISSN 1672-6316

Jibson, R., Harp, E. & Michael, J. (2000). A method for producing digital probabilistic seismic landslide hazard maps. *Engineering Geology*, Vol. 58, No. 3-4, (December 2000),pp. 271-289, ISSN 0013-7952

Keefer, D. (1984). Landslides caused by earthquakes. *Geological Society of America Bulletin*, Vol. 95, No. 4, (April 1984), pp. 406-421, ISSN 0016-7606

Keefer, D. (2000). Statistical analysis of an earthquake-induced landslide distributionthe 1989 Loma Prieta, California event. *Engineering Geology*, Vol. 58, No. 3, (December 2000), pp. 231–249, ISSN 0013-7952

Keefer, D. (2002). Investigating landslides caused by earthquakes – A historical review. *Surveys in Geophysics*, Vol. 23, No. 6. (November 2002), pp. 473-510, ISSN 0169-3298

Kerle, N., de Leeuw, J., (2009). Reviving legacy population maps with object-oriented image processing techniques. *IEEE Transactions on Geoscience and Remote Sensing*, Vol. 47, pp. 2392-2402

Khattak G A, Owen L A, Kamp U, Harp E L. (2010). Evolution of earthquake-triggered landslides in the. Kashmir Himalaya, northern Pakistan. *Geomorphology*, Vol. 115, pp. 102–108

Khazai, B., Sitar, N., (2004). Evaluation of factors controlling earthquake-induced landslides caused by Chi-Chi earthquake and comparison with the Northridge and Loma Prieta events. *Engineering Geology*, Vol. 71, No. 1, (January 2004), pp. 79–95, ISSN 0013-7952

Lee, C., Huang, C., Lee, J., Pan, K., Lin, M., Dong, J. (2008). Statistical approach to earthquake-induced landslide susceptibility. *Engineering Geology*, Vol. 100, No. 1, (June 2008),pp. 43–58, ISSN 0013-7952

Lin, C.W., Liu, S.H., Lee, S.Y., Liu, C.C., (2006). Impacts of the Chi-Chi earthquake on subsequent rainfall induced landslides in central Taiwan. *Engineering Geology* 86, pp. 87-101

Lin C W, Shieh C L, Yuan B D. (2003). Impact of Chi-Chi earthquake on the occurrence of landslides and debris flows: example from the Chenyulan River watershed, Nantou, Taiwan. *Engineering Geology*, Vol. 71, pp.49–61

Miles, B. & Keefer, D. (2009). Evaluation of CAMEL - comprehensive areal model of earthquake-induced landslides. *Engineering Geology*, Vol. 104, No. 1 , (February 2009), pp. 1–15, ISSN 0013-7952

Nefeslioglu, H.A., Gokceoglu, C., Sonmez, H. (2008). An assessment on the use of logistic regression and artificial neural networks with different sampling strategies for the preparation of landslide susceptibility maps. *Engineering Geology*, Vol. 97, pp. 171 – 191

Papadopoulos, G. & Plessa, A. (2000). Magnitude–distance relations for earthquake-induced landslides in Greece. *Engineering Geology*, Vol. 58, No. 4, (December 2000), pp. 377–386, ISSN 0013-7952

Pradhan.B, S.Lee. (2010). Landslide susceptibility assessment and factor effect analysis: backpropagation artificial neural networks and their comparison with frequency ratio and bivariate logistic regression modeling. *Environmental Modelling & Software*, Vol. 25, pp. 747 – 759

Qi, S., Xu, Q., Lan, H., Zhang, B. & Liu, J. (2010). Spatial distribution analysis of landslides triggered by 2008.5.12 Wenchuan Earthquake, China. *Engineering Geology*, Vol. 116, No. 2, (August 2010), pp. 95-108, ISSN 0013-7952

Qi, S., Xu, Q., Zhang, B., Zhou, Y., Lan, H. & Li, L. (2011). Source characteristics of long runout rock avalanches triggered by the 2008 Wenchuan earthquake, China. *Journal of Asian Earth Sciences*, Vol. 40, pp: 896-906, ISSN 1367-9120

Rodriguez, C., Bommer, J. & Chandler, R. (1999). Earthquake-induced landslides: 1980-1997. *Soil Dynamic and Earthquake Engineering*, Vol. 18, No. 5, (March 1999), pp. 325-346, ISSN 0267-7261

S.Lee, D. G. Evangelista. (2006). Earthquake-induced landslide-susceptibility mapping using an artificial neural network", *Nat. Hazards Earth Syst. Sci.*, Vol. 6, pp. 687–695

Shaller, PJ. & Shaller, AS. (2006). Review of proposed mechanisms for sturzstroms (long runout landslides). *Sturzstroms and Detachment Faults*, Anza-Borrego State Park, California, pp.185-202

Tang C, Zhu J, Li W L. (2009) Rainfall triggered debris flows after Wenchuan earthquake. *Bull Eng Geol Environ*, Vol. 68, pp. 187–194

Tang C, Ding J, Qi X. (2010). Remote sensing dynamic analysis of rainstorm landslide activity in Wenchuan high-intensity earthquake area. *China university of geosciences journal*, Vol.35, pp. 317-323

Tang C, Zhu J, Qi X. (2011a). Landslide Hazard Assessment of the 2008 Wenchuan Earthquake: a case study in Beichuan Area. *Canadian Geotechnical Journal*. Vol. 48, pp. 128–145

Tang, C., Zhu, J., Qi, X. & Ding, J. (2011b). Landslides induced by the Wenchuan earthquake and the subsequent strong rainfall event: A case study in the Beichuan area of China. *Engineering Geology*, doi:10.1016/j.enggeo.2011.03.013, ISSN 0013-7952

Tapas R. Martha, Norman Kerle, Victor Jetten, Cees J. van Westen, K. Vinod Kumar. (2010). Characterising spectral, spatial and morphometric properties of landslides for semi-automatic detection using object-oriented methods. *Geomorphology* Vol. 116, PP. 24-36

Wu, S., Wang, T., Shi, L., Sun, P., Shi, J., Li, B., Xin, P. & Wang, H. (2010). Study on catastrophic landlsides triggered by 2008 great Wenchuan earthquake, Sichuan, China. *Journal of Engineering Geology*, Vol. 18, No. 2, pp. 145-159, ISSN 1004-9665 (in chinese)

Xie, H., Zhong, D.L., Jiao, Z., et al.,2008.Debrisflow in Wenchuan quake-hit area in 2008.Journal of Mountain Science 27(4),501-509.

Xu, Q., Pei, X., Huang, R. et al. (2009a). *Large-scale Landslides Induced by the Wenchuan earthquake*, Science Press, ISBN 978-7-03-026906-5, Beijing, China (in chinese)

Xu, Q., Fan, X., Huang, R. & Westen, C. (2009b). Landslide dams triggered by the Wenchuan earthquake, Sichuan Province, south west China. *Bulletin of Engineering Geology and the Environment*, Vol. 68, No. 3, (August 2009), pp. 373-386, ISSN 1435-9529

Yagi, H., Sato, G., Higaki, D., Yamamoto, M. & Yamasaki, T. (2009). Distribution and characteristics of landslides induced by the Iwate-Miyagi Nairiku Earthquake in 2008 in Tohoku District, Northeast Japan. *Landslides*, Vol. 6, No. 4, (December 2009), pp. 335–344, ISSN 1612-510X

Yin, J., Chen, J., Xu, X., Wang, X. & Zheng, Y. (2010). The characteristics of the landslides triggered by the Wenchuan Ms 8.0 earthquake from Anxian to Beichuan. *Journal of Asian Earth Sciences*, Vol. 37, No. 6, (March 2010), pp. 452-459, ISSN 1367-9120

Yin, Y., Wang, F. & Sun, P. (2009a). Landslide hazards triggered by the 2008 Wenchuan earthquake, Sichuan, China. *Landslides*, Vol. 6, No. 2, (June 2009), pp. 139-151, ISSN 1612-510X

Yin, Y. (2009b). Rapid and long run-out features of landslides triggered by the Wenchuan Earthquake. *Journal. of Engineering. Geology*, Vol. 17, pp. 153–166 (in Chinese)

7

Experience with Restoration of Asia Pacific Network Failures from Taiwan Earthquake

Yasuichi Kitamura[1], Youngseok Lee[2],
Ryo Sakiyama[3] and Koji Okamura[3]
[1]National Institute of Information and Communications Technology,
[2]Chungnam National Univeristy,
[3]Kyushu University
[1,3]Japan
[2]Korea

1. Introduction

As the Internet grows, networks become larger and more complex, and the number of components, such as routers, switches, and fiber cables, increases. In complicated network systems, it is difficult to implement global network management across several Internet service providers (ISPs) that use a lot of network components in a large-scale network topology. Fault management is a particularly important network management issue in complex network systems because the Internet has become essential to business and research. However, we are only beginning to learn how to deal with global network failures in large networks.

Failures have been reported in Sprint Internet protocol (IP) backbone, which shows that failures can be observed in everyday operation (Iannanccone et al., 2002; Markopulou et al., 2004). However, the network failures observed by (Iannaccone et al., 2002) and (Markopoulou et al., 2004) were short-lived and small scale, and their impacts were analyzed only in the context of a single ISP. Most network backup or fault restoration methods have been studied and proposed for the various layers such as wavelength division multiplexing (WDM), multi-protocol label switching (MPLS), or IP (Fumagalli & Valcarenghi, 2000; Gerstel & Ramaswami, 2000; Ramamurthy et al, 2003; Saharabuddhe et al., 2004; Sharma & Hellstrand, 2003). Yet, the proposed backup and restoration methods have not been fully implemented and deployed in the real network. Since real networks are more complicated than theoretical ones, the impacts of network failures on users and ISP's cannot be completely predicted and analyzed. Significant network failures due to natural disasters such as earthquakes, floods, or fires could have particularly wide impact on several ISPs.

We discuss the results of the critical network failures that occurred after the Taiwan earthquake in Dec. 2006, which cut fibers and caused network failures. We also explain how restoration methods such as automatic border gateway protocol (BGP) (Lougheed & Rekhter, 1989) re-routing, BGP policy change, and switch reconfiguration were conducted. We hope that the experience and knowledge we gained during the process of recovering from this huge natural disaster, which affected the global Internet, can be shared and can contribute to future Internet network management research. To the best of our knowledge, this is the first detailed study of network restoration after global network failures due to a natural disaster.

Although many natural disasters have occurred in the 21st century, until recently there had been no simultaneous outage of the global Internet backbone. However, the earthquake that occurred around Taiwan in 2006 made several Asia Pacific Research and Education (R&E) networks unreachable. At 21:26 and 21:34 on December 26th (UTC+9), 2006, there was a big undersea earthquake off the coast of Taiwan twice, which measured 7.1 and 6.9 respectively on the moment magnitude (Hanks & Kanamori, 1979). This earthquake caused significant damage to the undersea fiber cable systems in that area. Several ISPs were affected because each cable system is shared by multiple ISP's. This earthquake had the effect of dividing the Asia Pacific R&E networks into an eastern and a western group. The Asia Pacific R&E networks were, in particular, seriously damaged but were fully restored after several restoration steps, including automatic BGP re-routing, BGP policy changes, and switch port reconfigurations, were taken.

The first step in recovery after the earthquake was taken automatically by BGP routers, which detoured traffic along redundant routes. In BGP routing, there are usually multiple redundant AS paths. Redundant BGP routes have served as backup paths but have provided poor quality connectivity, i.e., long round trip time (RTT). Because of the congestion on the narrow-bandwidth link that was subsequently reported, operators took manual control of traffic to improve communication quality. The second step was a traffic engineering process intended to prevent narrow-bandwidth links from filling up with detoured traffic. The operators changed the BGP routing policy related to the congested ASs. In spite of the routing-level restoration, a few institutions were still not directly connected to the R&E network community because they had only a single link to the network. For these single-link networks, the commodity link was used temporarily for connectivity. However, the commodity link was not stable and not sufficient to carry a huge amount of bandwidth or to provide next generation Internet service. To restore the single-link networks, cable connection configurations at the switches were changed.

The fiber break caused by the Taiwan earthquake raised restoration issues related to BGP re-routing. In such an emergency, the backup routes should be chosen based on available bandwidth and RTT. Since the fiber break required an urgent network recovery process, network operators configured re-routing based on their experience with bandwidth and RTT.

From this experience, we have learned that redundant physical backup links and routes are important to providing bandwidth and connectivity and that the Quality-of-Service (QoS) after recovery is also important. From the viewpoint of restoration after network failures, there are still challenges that cannot be automatically overcome by network management systems. A systematic risk management plan that includes collaboration among operators of the next-generation Internet is needed.

The remainder of this chapter is as follows. In Section 2, the Asia Pacific R&E networks that were damaged by the earthquake or related events are introduced. Section 3 introduces the R&E connection especially in Asia Paicifc area and the issues caused by such inter-connectivity of R&E networks. Section 4 is a detailed report of the network failures that were observed after the earthquake. Section 5 describes the processes to restoring the disrupted communications in the area. Section 6 discusses what we have learned from the observation of the network failures and recovery processes. Finally, we conclude the paper in Section 7.

2. Research and education network activities in the Asia Pacific area

2.1 Asian Internet Interconnection Initiative (AIII)

When AIII (Asian Internet Interconnection Initiative [AIII], n.d.) started in 1996, this project was the first next generation Internet R&E network in Asia. The basic idea of AIII is to provide Internet services via satellite. There are AIII members in TH, MY, ID, and SG (Tbl. 3). NP also recently joined the project. The AIII's major effort is to build up the access points of non-broadband networks. In Asia, it was very hard to complete the network over telephones lines, and preparing earth stations represented a better chance of accessing the Internet. The biggest drawback of this project is that the total bandwidth is limited to between 1.5 and 8 Mbps. In 1996, 2 Mbps was enough bandwidth to start research activities. Now, however, even 8 Mbps is insufficient for network technology research activities. These days, AIII concentrates on developing and deploying certain advanced technology on their network, for example, IPv6 unicast, IPv6 multicast, Uni Directional Link Routing (UDLR), and advanced TCP.

2.2 Asia-Pacific Advanced Network (APAN)

APAN (Asia-Pacific Advanced Network [APAN], n.d.) started in 1997. APAN is the research consortium as well as it operates a next generation Internet service. The bandwidths of the backbone networks of APAN in 1998 were between 1 Mbps and 35 Mbps. The US next generation Internet project "very high speed Backbone Network Service" (vBNS) already had 155 Mbps bandwidth service in their backbone network. The operating policy of APAN was to provide high performance data transfer service, because, in 1998, it was still impossible to implement a huge bandwidth network in the Asia Pacific area. Now, the APAN network covers the Asia Pacific area, providing bandwidth between 45 Mbps and 10 Gbps. In 2011, APAN became the none-profit organization and APAN strongly supports the research activities over the R&E networks in Asia Pacific area.

2.3 Trans-Eurasian Information Network 2 (TEIN2)

TEIN or TEIN1 was an EU project that connected Europe and Korea. It started with a bandwidth of 10 Mbps. TEIN2 (Delivery of Advanced Network Technology to Europe [DANTE], n.d.) is a bit different from TEIN in that it has two goals. One is to provide an access network to Europe, and the other is to develop an interconnection network in the Southeast Asian area whose bandwidth is between 45 Mbps and 1 Gbps. The main characteristic of the design topology is that it is not the star-shaped. That is, the TEIN2 network itself has its own backbone with connecting the four NOCs, and the number of routes for communicating between NOCs are more than two. It provides each Southeast Asian site with several routes to access the others. TEIN3 started in 2010 and TW, KH, LA, IN, LK, NP, PK, BD and BT (Tbl. 3) joined this activity.

3. Background of R&E network

3.1 R&E network status in 2006

In the old star-shaped R&E network topology (Fig. 1) communication between the point sites had to go through the center of the star. Communication along these routes was frequently delayed due to the configuration. Now, however, the R&E network community has grown, and the former point sites are now sometimes the center of a star. In some cases this growth generated several routes between two sites (Fig. 2).

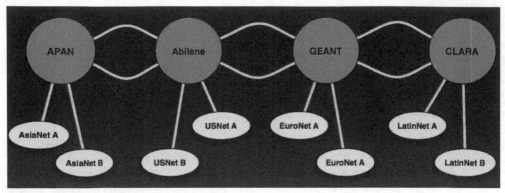

Fig. 1. Star shaped R&E networks(Robb, 2006)

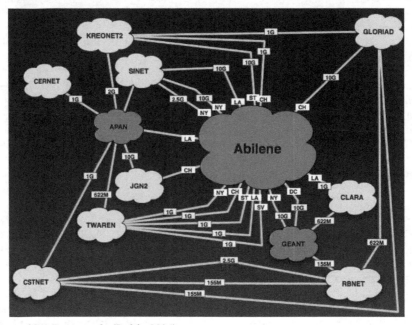

Fig. 2. Actual R&E networks(Robb, 2006)

This complicated topology made the operation of the R&E networks more difficult. To simplify operations, Research and Education Network Operators Group (RENOG), (Research and Education Network Operators Group [RENOG], n.d.) was developed and started reconfiguring the complicated and unstable routing. However, in the APAN area, the complicated routing unexpectedly worked and it was able to maintain high-speed communication with most of the network researchers, though with some delay. Network engineers like the word "redundancy" but most of the network engineers in the APAN area did not expect that the complicated topology and the complicated routing information would work so well in an emergency.

3.2 Before the earthquake

The TEIN2 project started in 2006. Before that, the network topology style in Asia Pacific area was close to the star shape. Most of the networks in the APAN area started in Japan, and if a network did not start from Japan, it started from one of the point sites. That is why the star shape was kept. However, the TEIN2 topology does not look like a star. Even inside TEIN2, there are two or three routes to reach any other sites. An engineering meeting was held before TEIN2 started at which an agreement on routing policy seemed to have been reached. However, after TEIN2 started, there happened some of routing troubles in Asian Pacific area. A few troubles were from TEIN2 network directly but others were the routing advertise issues from the National Research and Education Networks (NRENs) of the TEIN2. To solve these problems, the operators held many meetings to develop tools. One tool summarized traffic conditions for each route (Asia-Pacific Advanced Network Japan [APAN-JP], n.d.). Another was a database to record preferable routes for each XP (Kurokawa, 2006). These were essentially monitoring and advising tools, not route configuration tools. Actual operations were done at the XPs based on the database, which was built by the database tool (Kurokawa, 2006). At the APAN Tokyo XP, the abstract of the routing policy was as follows.

- Communication with ID, SG, MY, TH, VN, CN, and AU (Tbl. 3) should be routed through TEIN2.
- Communication with PH should be direct.
- Communication with Micronesia should be routed through the Hawaii XP.
- Communication with TW should be direct.
- Generic communication with EU should be through US.
- Communication with US should be direct.

Figure 3 shows the one of the worst examples of communication between US university through KR. That is, if a university in US established communication with Merit or the University of Michigan, the packets were transferred through KR. In such a case, TCP-based applications often met with communication problems.

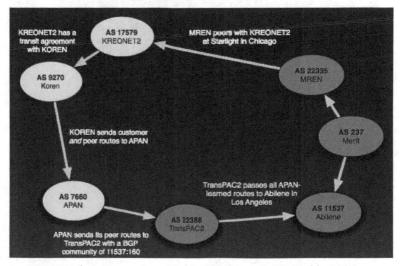

Fig. 3. Incorrect routing between universities in US through KR(Robb, 2006)

4. Network failures caused by Taiwan earthquake

4.1 Fiber breaks

On December 26th, 2006 (UTC+9) there were two huge earthquakes near Taiwan. The first earthquake happened at 20:26 (UTC+8) (Academia Sinica, n.d.a), and the second one at 20:34 (UTC+8) (Academia Sinica, n.d.b). Fortunately, the earthquakes took place under the sea and the cities in TW were not heavily damaged, as happened in 1999. However, these two earthquakes did cause landslides over a wide area on the seabed near the Taiwan island. At 04:00 (UTC+9) on December 27th 2006, that is, after the second earthquake, the R&E networks in the Asian area were shutdown. The cable companies investigated the reason for the lost connection and found that the earthquake had caused damage to the cable systems.

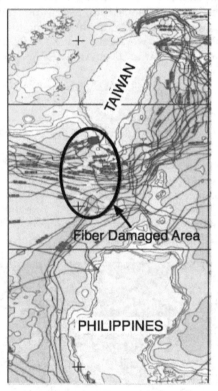

Fig. 4. Fiber cut area around Taiwan island (Konishi, 2007)

The circle in Fig. 4 shows the area where the cable systems were cut off. Most of the fiber cables in the eastern Asia area went through southwestern Taiwan. These cables were generally bought and shared by different telecom companies.

4.2 Internet disconnections and lost BGP peerings

After the earthquake, both commodity Internet traffic and R&E traffic were cut off. For instance, the JP-PH, JP-CN, JP-SG, CN-US, HK-KR, TW-(HK+CN), and TW-SG connections were lost. That is, the R&E network communities were divided into two groups. One was the group that consisted of JP, KR, TW, and US, and the other consisted of CN, HK, VN, MY, TH, SG, ID, and PH (Fig 5).

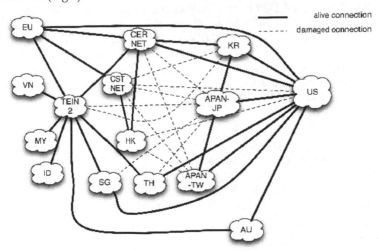

Fig. 5. Splitted R&E networks in Asia Pacific area

The Internet disconnection occurred in the following order.
1.   Link-layer disconnection because of the fiber cut
2.   Lost primary BGP peerings
3.   Automatic BGP re-routing along the alternative peer if any

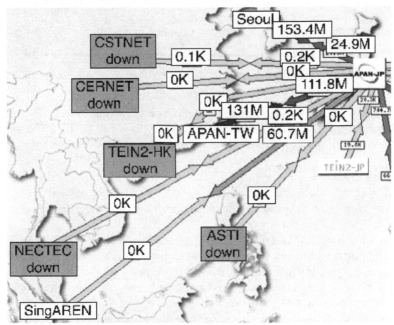

Fig. 6. Traffic weather map on Dec.27 2006

Figure 6 shows the Internet traffic weather map at APAN Tokyo XP (APAN-JP, n.d.), which displays connectivity and link utilization in real time. In Fig. 6 it can be observed that the JP-HK-CN, JP-TH, JP-SG, and JP-PH communications were lost and that there was 0% link utilization except a 100 Mbps load between JP and KR.

The link-layer disconnection caused BGP sessions to expire. BGP peerings from JP to HK+CN, TH, SG, and PH were lost and automatically diverted to detour routes.

In general, when traffic is transferred to the detour AS routes, the traffic will flow along the longer AS path rather than the usual one, because the shortest AS path will be selected as the primary AS path according to BGP policy.

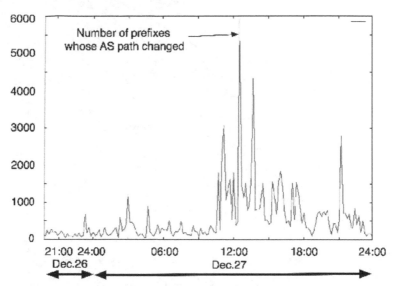

Fig. 7. Number of IP prefixes that experienced AS path changes (2006.12.26 - 2006.12.27 at QGPOP)

Figure 7 shows the AS path changes of each IP prefix observed just after earthquake from the QGPOP (Kyushu GigaPOP Project [QGPOP], n.d.) BGP router in JP. It can be seen in Fig. 7 that more than 1000 IP prefixes experienced AS path changes after the earthquake.

4.3 Traffic load changes

Fortunately, despite the earthquake, the CN-KR and KR-JP cables were unbroken. Therefore, we were able to observe the detour traffic along these links due to BGP re-routing.

Figure 8 shows the traffic between JP and KR on Dec. 27th, 2006. At about 04:30 (UTC+9) the inbound traffic pattern between JP and KR had changed dramatically. At the same time, as can be seen in Fig. 9, the traffic between JP and PH disappeared.

The reason for the traffic change could be inferred from the routing policy of APAN Tokyo XP. The route from CN to JP through KR was one of the lowest priority routes, but after the earthquake, it was chosen because there were no available BGP routes with high priorities.

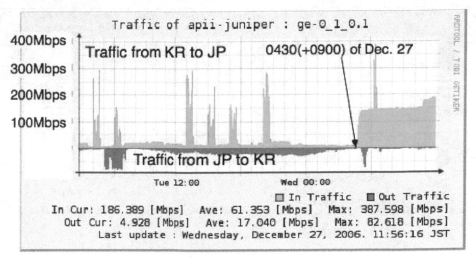

Fig. 8. Traffic between JP and KR on Dec.27 (UTC+9)}

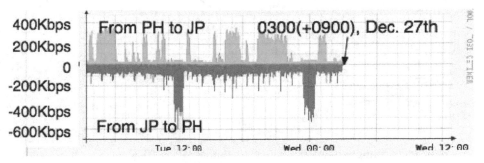

Fig. 9. Traffic between JP and PH on Dec.27 (UTC+9)

4.4 Changes of BGP routing tables

Table 1 shows the registered routing policy table of the Asia Pacific R&E networks. It can be observed that ASTI (PH) lost connectivity to the R&E networks, that CSTNET (CN) lost eastbound routes, and that APAN-JP (JP) lost connectivity to TEIN2.

CERNET (CN) had a direct connection to TW, but after the earthquake its connection was expected to be changed to the path through US, as shown in Fig. 10. But, the routing policy arrangement was different between JP and KR. The forward path between CN and TW chose the route through JP but the return path chose the route through US as shown in Fig. 11.

In addition, routing from JP to the "west" Asian networks (Fig. 5) was connected through US. The direct link between CSTNET (CN) and US was also damaged.

Both CERNET and APAN Tokyo XP expect that the detour for JP-CN traffic should be through US, not through KR. Figure 10 shows the expected BGP route and the actual route between APAN-TW and CN. Since only APAN Tokyo XP implemented the strict routing policy, the CN traffic chose the shortest AS path. However, the traffic from JP to CN chose a routing policy that does not choose the route through KR.

Src\Dst	APAN-JP (JP)	CERNET (CN)	APAN-TW (TW)	KOREN (KR)	ASTI (PH)	CSTNET (CN)	TEIN2	SingAREN (SG)	UniNet (TH)	ThaiSARN (TH)	KREONET2 (KR)	AARNET (AU)
APAN-JP		D	direct	direct	U	C	U	direct	D	U	US	US Hawaii
CERNET	D		D	direct	U	N.A.	TEIN2	TEIN2	TEIN2	TEIN2	TEIN2	TEIN2
APAN-TW	direct	D		APAN-JP	U	U	N.A.	N.A.	N.A.	N.A.	APAN-JP	APAN-JP
KOREN	direct	direct	APAN-JP		U	U	TEIN2	TEIN2	TEIN2	TEIN2	direct	TEIN2
ASTI	U	U	U	U		U	U	U	U	U	U	U
CSTNET	C	N.A.	U	U	U		N.A.	N.A.	N.A.	N.A.	U	N.A.
TEIN2	U	TEIN2	N.A.	TEIN2	U	N.A.		TEIN2	TEIN2	TEIN2	KOREN	TEIN2

D:	Detoured path through US	U:	unreachable
C:	Commodity link connection	N.A.	Non applicable

Table 1. The monitored routing table for each R&E network on Dec.27 2006 (Kurokawa, 2006)

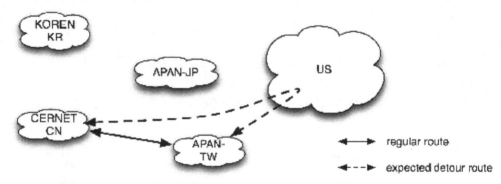

Fig. 10. BGP route changes of APAN before earthquake (Dec. 2006)

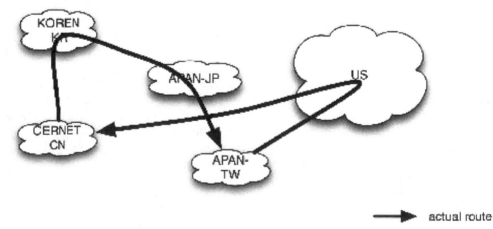

Fig. 11. BGP route changes of APAN after earthquake (Dec. 2006)

4.5 AS-level topology changes

To investigate the BGP route changes in detail, we used an AS-topology visualization tool called "ABEL2" (Sakiyama et al., 2006) that utilizes BGP routing tables that are stored every 10 minutes.

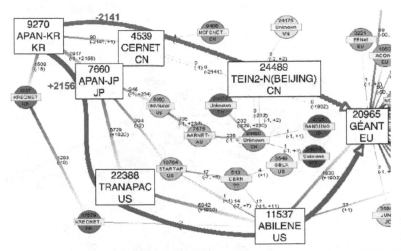

Fig. 12. After the earthquake at 19:30, Dec. 27 (UTC+9) 2006, the route from APAN-KR to GÉANT was diverted to a long AS path of (APAN-KR, APAN-JP, TransPAC, Abilene, GÉANT)

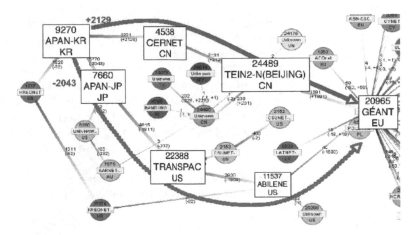

Fig. 13. Reconfigured AS path from APAN-KR to GÉANT (APAN-KR, CERNET, TEIN2-N, GÉANT) at 20:30, Dec. 27 (UTC+9) 2006

Figures 12, 13, and 14 show the changes between APAN-KR (AS9270) and GÉANT (AS20965) by the number of IP prefixes. It can be observed in Fig. 12, that just after the earthquake, at 19:30 on December 27th (UTC+9), 2006, the route from APAN-KR to GÉANT was diverted to the long AS path APAN-KR - APAN-JP - TransPAC (US) -

Abilene (US) – GÉANT (Tbl. 4) because of the TEIN2 link outage. Therefore, to connect APAN-KR to GÉANT with a shorter AS path, the operator configured the BGP routing policy to make CERNET (CN) announce GÉANT prefixes. As shown in Fig. 13, the route from APAN-KR to GÉANT was switched through CERNET at 20:30 on Dec. 27th (UTC+9), 2006. At 20:50 on Dec. 17th, 2006 (UTC+9), the TEIN2-SG NOC announced the route to EU, too. Since the link bandwidth between SG and KR is larger than that between CN and KR, the operator made a configuration for BGP routers to choose the AS path with the KR-SG link (Fig. 14).

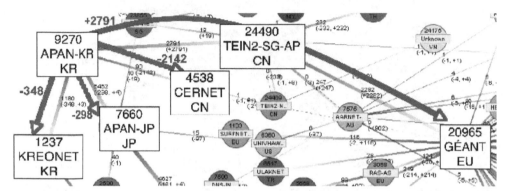

Fig. 14. Reconfigured AS path from APAN-KR to GÉANT through TEIN2-SG at 20:50, Dec. 27 (UTC+9) 2006

4.6 Delay changes

When Australia's Research and Education Network (AARNET) proposed that it would transit all the traffic of TEIN2 members (CN, SG, VN, MY, TH, ID), a flow data analysis was done and most of the traffic was found to be for Eastern Asia.

The APAN Tokyo XP and the TEIN2 NOC asked the AARNET NOC to route the TEIN2 traffic through Honolulu rather than through Seattle, because the traffic was between the Eastern Asia and TEIN2 members. The shorter Round Trip Time (RTT) worked better, especially with TCP-based applications (Tbl. 5). However, this operation occupied the link between Honolulu and Tokyo, because the bandwidth between Tokyo and Honolulu is 155 Mbps and the one between the Tokyo and TEIN2 NOCs is 622 Mbps and the one between AARNET and TEIN2 is 1 Gbps (Fig. 15).

The link between KR and JP, whose bandwidth was 10 Gbps, was up. According to the analysis, the KR-JP link still had space to carry the traffic, and this route is much shorter than the route through Australia.

The TEIN2 and AARNET NOCs and the Tokyo XP stopped routing the TEIN2 traffic through Australia, and the KR transit policy was initiated. But the bandwidth between CN and KR was also 155 Mbps, too, so the link was occupied again (Fig. 16).

At that moment, the JGNII JP-SG link (Tbl. 4) was restored, but after precise check by the engineers, it looked as though the telecommunication company had switched the route and provided connectivity. TEIN2 NOC and the Tokyo XP decided to separate the traffic to Tokyo into two routes. That is, CN traffic was transferred through KR, and other TEIN2 traffic was transferred through SG (Fig. 17).

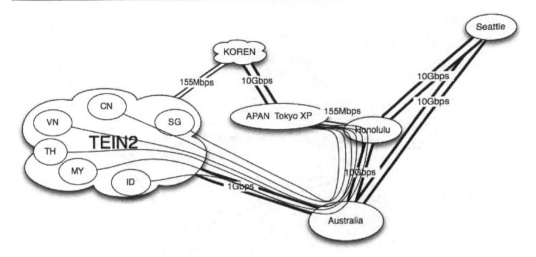

Fig. 15. TEIN2 traffic was carried through Hawaii.

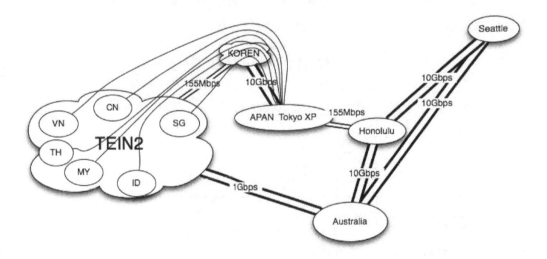

Fig. 16. TEIN2 traffic was routed through KR.

On December 27th after the fiber cut, automatic BGP re-routing has been carried out between SG and JP. The route from SG to JP became SG-AU-Hawaii-JP instead of the direct link and its RTT was increased to 426 ms, while its normal RTT is around 88 ms.

When the link between SG and KR was temporarily recovered with the backup fiber, the RTT values was 240 ms between SG and JP via KR on December 28th.

Finally, on January 12th when the link between JP and SG was recovered with the direct fiber, the RTT was reduced to 113 ms, which is slightly increased than the usual case.

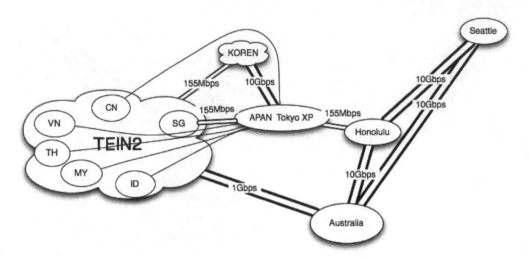

Fig. 17. TEIN2 traffic was routed through SG and KR.

Figure 18 shows the RTT between SG and JP just after the earthquake and the recovery process.

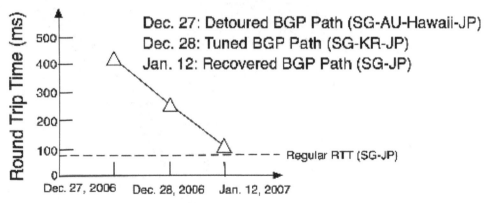

Fig. 18. RTT between SG and JP during the restoration process

5. Network restoration methods

After the network failures caused by the earthquake, several restoration steps were taken to restore communication. In this subsection, we discuss these steps.

5.1 Automatic BGP re-routing

Usually, the full BGP routing table includes a few "useless" routes (Tbl. 2). By "useless" we mean that the route itself provides only connectivity with the long RTT and insufficient bandwidth. Therefore, the network operators filter out such useless routes by setting the local preference to ignore them. However, after the earthquake, these useless BGP routes worked automatically as backup paths. In the Asia Pacific R&E networks, the routes became very complicated after TEIN2 started because TEIN2 provided a few unexpected routes around the world. Because there were backup AS paths, automatic BGP re-routing could be used for first aid to provide the connectivity to the ASs that lost the primary paths. However, automatic BGP re-routing did not consider the traffic engineering parameters of the available bandwidth and the backup traffic load.

(Src, Dst)	Usual Path	Useless Path
(JP, TEIN2)	JP-TEIN2	JP-Hawaii/Seattle-AU-TEIN2
(TEIN2, JP)	TEIN2-JP	TEIN2-GÉANT2-Abilene-JP
(US, US)	Abilene	US-KREONET2-JP-TransPAC2-US
(JP, KR)	JP-KR link	JP-TEIN2-JP- TEIN2-SG -AU-Seattle-KR

Table 2. Examples of the "useless routes"

5.2 Traffic engineering with BGP policy change
BGP by itself does not provide any information regarding link capacity or available bandwidth. Moreover, due to recent VLAN (Varadaraja, 1997) technology, the distance between two ASs has no relation to physical distance. Thus, QoS information of the detour routes must be examined by the operators. This makes systems reliant on human knowledge of traffic engineering. To remove the congestion due to the long detour AS path, we changed the BGP routing policy as shown in Fig. 13 and Fig. 14.
The members of TEIN2 (VN, MY, SG, ID, PH) lost their connections to APAN Tokyo XP because the fiber broke. AARNET NOC proposed backup routes for accessing APAN Tokyo XP through AU and Hawaii. However, this solution caused congestion on both the CN-KR and Hawaii-JP links. Besides, CN traffic took an asymmetrical path.
Therefore, to solve the traffic engineering issue, Tokyo XP made a decision to divide CN traffic by announcing CN IP prefixes through KR NOC and grouping CN prefixes at Tokyo XP. The results were monitored by Cisco NetFlow (CISCO, n.d.). The operators found out that half of the KR traffic was from CN. After a careful examination, it was discovered that a part of the CN traffic was from CERNET but the other part was from TEIN2.

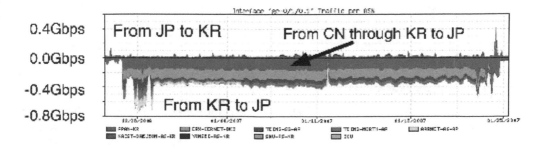

Fig. 19. Traffic from KR monitored at APAN JP (Dec. 26 2006 - Jan. 26 2007)}

Figure 19 shows the traffic load for each source AS. Although the total traffic is about 0.4 Gbps, the real KR traffic was about 0.2 Gbps. 0.1 Gbps is occupied by CERNET traffic and the rest by TEIN2.

5.3 Port reconfiguration
In spite of the recovery steps, a few sites that had only single-link connections to the Internet could not directly reach the R&E networks. Therefore, after the fiber was fixed, a few single-link sites had the connections via Internet commodity service. To fix this problem, the operators had to change the port configuration at switches that were able to provide high performance connections.

Two days after the earthquake, the link between KR and SG was restored and SG started making routing announcements to TEIN2 members (ID, SG, TH, MY). However, the link between SG and KR was not the original fiber, and its RTT increased greatly because a detour route was assigned. Similarly, the direct link between PH and JP was replaced with a detour route with a long RTT. When an online demonstration was being prepared for the 2007 APAN meeting held in PH between January 22nd and 26th 2007, the restored JP-PH link had a very long RTT because it went through mainland China, as shown in Fig. 20.

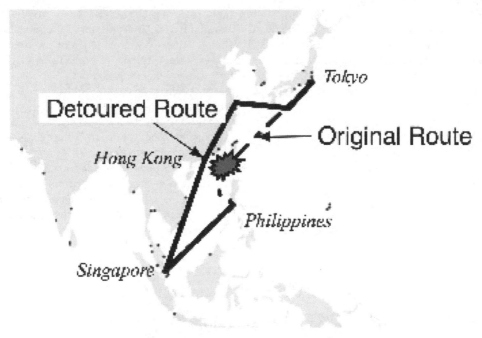

Fig. 20. JP-PH detour route circuit

During the 23rd APAN meeting demonstration, the Prince of Wales Hospital at the Chinese University of Hong Kong (CUHK) was expected to join. However, its IP prefix was announced only by CSTNET. Thus, it was only reachable over a commodity Internet link with a small bandwidth. To solve this problem, APAN Tokyo XP changed the port configuration of fibers. That is, the CSTNET fiber for accessing APAN Tokyo XP was plugged to the TEIN2-HK router. Finally, CUHK was directly connected to TEIN2. Finally, with this solution, CUHK and CSTNET were supplied with huge-bandwidth and short-RTT connections to the R&E network.

6. Lessons

From the process of recovering from network failures across several ISP's in Asia Pacific R&E networks, we encountered several network management challenges especially regarding fault management. In this subsection, we describe the lessons learned during the recovery operations.

6.1 Fault-tolerant fiber topology design

BGP routing policies are usually made to avoid asymmetric or useless AS paths by setting the appropriate local preference values. However, these alternative AS paths worked as backup paths. Before the network failures from earthquake, Asia Pacific R&E network operators thought that removing the useless routes was urgent, because routing became too complicated after TEIN2 started. However, this complicated routing was able to provide valuable connections during network failures. This shows that maintaining full-mesh style routing information is very important for fault-tolerant routing.

Though BGP re-routing over the redundant AS paths was successful for the first step in restoration, it was not sufficient to provide full backup service without congestion by considering the traffic load. Since BGP routing does not carry QoS information, such as link capacity, link utilization, or available bandwidth, traffic re-routed to the backup AS path had experienced poor QoS, such as long delays. Therefore, QoS-aware BGP routing or traffic engineering-aware BGP routing is necessary.

6.2 Integrated network management

During the restoration process, we used various network monitoring tools such as an MRTG (Oetiker, n.d.a), a network weather map, a BGP routing table visualizer, and a flow monitor. At first, the link outage was noticed on the network weather map, and the abrupt change of traffic load was noticed on the MRTG. However, the fast fault detection method that encompasses physical, link, routing, and application layers is necessary because it was able to identify the exact failure points and visualize their impacts on the network. In addition, a simulator or emulator that could show the results with the network topology and the traffic load before and after failures would be very useful in predicting the effects of fault-management decisions. While we took various restoration steps, we had to process the information collected by each different network-monitoring tool. Finally, the operators interpreted the situation and implemented recovery decisions manually. If the iperf (GOOGLE, n.d) or bwctl (INTERNET2, n.d.) is available throughout the network, the end-to-end available bandwidth between ASs can be easily estimated. For example, to access Sydney from Tokyo, there are two possible routes. One is Tokyo – Seattle - Sydney, and the other is Tokyo – Honolulu - Sydney. The former provides 10 Gbps but has a long RTT. The latter route includes a bottleneck along the 155 Mbps path but has a short RTT. In addition, to make the final decision, we had to check the flow data, because MRTG or RRDTool (Oetiker, n.d.b) do not classify traffic breakdowns by their source/destination ASs. When the traffic from KR increased suddenly, the operators could not understand the reasons. This shows that integrated network monitoring or management systems would be very useful for collecting information from several independent monitoring systems and for providing the correct information in an integrated wide view in case of significant network failures.

6.3 Emergency communication between operators

After the earthquake, communication among NOCs was difficult because the fiber break disrupted VoIP and legacy telephone service. Moreover, the earthquake happened on December 26th 2006, overlapping with the Christmas holiday. Thus, all the communication was routed over the instant messaging system and e-mails were routed over the detoured network even though it provided poor quality service. It became obvious that the emergency communication should be guaranteed in case of failures so that the recovery process can be started quickly

7. Summary

Since the Internet continues to grow globally and becomes ever more important in daily life, business, and research, the need for fault-tolerant service in network management becomes more urgent. However, during the network failures caused by the 2006 earthquake, it was shown that there are still many challenges in fault-tolerant network management research. Even though multiple fiber cores are installed together to provide backup service, they may be useless during severe natural disasters. Therefore, full-mesh or fiber-disjoint physical network topology should be designed for use during failures. On the available topology, it was seen that BGP routing provided backup AS paths, which was useful for the first step in restoration. However, the traffic engineering issues during restoration were difficult to solve because all the information, such as link capacity, available bandwidth, link delay, traffic load, and routing policy, had to be collected, interpreted, and acted on by human operators. In spite of BGP re-routing, we had to deal with a few single-link ASs to establish direct connections to the R&E networks. From this experience of network recovery during a significant natural disaster affecting several different countries and ISP's, we were able to gather valuable information on network management during emergencies. Therefore, in the Internet of the future, designers should focus on fault-tolerant network management study including robust physical topology, cross-layer restoration, traffic engineering combined with BGP routing, and simulation of failures in the network.

We would like to show the one example from this lesson. In August 2009, the earthquake happened again near Taiwan island and the fiber cut happened again. Table 1 tool was improved as "Compath" already and the table was constructed again for this disaster (Kurokawa, 2010). At that time, the medical demonstration was planned in Asian area and the connectivity of the Thailand was required. The compath table recommended that the TW should have become the hub for the southeast Asian area. This was against the policy of TEIN2, but the R&E networks object is to support the researchers and such the flexibility was approved and the tool from the lesson of the 2006 earthquake improved the network operations.

8. Acknowledgment

After the earthquake, KDDI OTC (Otemachi Technical Center) worked very hard to fix the routing problems of APAN Tokyo XP both in their capacity as paid staff and as volunteers.

Yutaka Watanabe, OTC's director, took the MRTG snapshot of Figs. 8 and 9. AARNET NOC offered us the backup routes to access the TEIN2. KOREN NOC worked very hard to keep communication open between CERNET and ASNET even during the holiday. Hawaii University NOC worked very hard to keep control of this complicated routing. KDDI investigated the reason for the communication failures in the APAN area and gave useful advice to APAN Tokyo XP. The staff of Genkai XP NOC accepted CERNET's traffic from KOREN by upgrading the JP-KR link bandwidth. Due to these great efforts and the collaboration among the network engineers, we were able to quickly restore the Asia Pacific R&E networks.

This research was supported by the MIC (Ministry of Information and Communication), Korea, under the ITRC (Information Technology Research Center) support program supervised by the IITA (Institute of Information Technology Assessment) (NIPA-2011-C1090-1131-0005).

9. Appendix

Country code	Name
JP	Japan
TH	Thailand
MY	Malaysia
HK	Hong Kong
ID	Indonesia
SG	Singapore
TW	Taiwan
CN	China
US	United States of America
KR	Korea
VN	Vietnam
PH	The Philipppines
NP	Nepal
AU	Australia
BD	Bangladesh
IN	India
LK	Sri Lanka
BT	Bhutan
KH	Cambodia
LA	Laos
PK	Pakistan

Table 3. ISO 3166 Country Code (International Organization for Standardization [ISO], n.d.) in this chapter

Network Name	Description	Area
AIII	Asian Internet Interconnection Initiatives	JP, VN, PH, TH, MY, SG, ID, NP
APAN	Asia-Pacific Advanced Network	JP, KR, CN, TW, HK, VN, TH, MY, SG, AU, NZ, PK, IN, KL, BD
TEIN2	Trans-Eurasia Information Network 2	JP, CN, VN, PH, TH, MY, SG, ID, AU, EU
TEIN3	Trans-Eurasia Information Network 3	AU, BD, BT, KH, CN, IN, ID, JP, KR, LA, MY, NP, PK, PH, SG, LK, TW, TH, VN
ASTI	Advanced Science and Technology Institute, the link owner of PREGINET	PH
PREGINET	Philippine Research Education and Government Information Network	PH
CSTNET	China Science & Tecnology Network	CN
CERNET	China Education and Research Network	CN
GÉANT	R&E Network, the AS registration name	Europe
GÉANT2	R&E Network	Europe
KOREN	Korea Advanced Research Network	KR
SingAREN	Singapore Advanced Research and Education Network	SG
UniNet	inter University Network	TH
ThaiSARN	Thai Social/Scientific Academic and Research Network	TH
KREONET2	Korea Research Environment Open Network 2, the domestic name is KREONET	KR
AARNET	Australia's Research and Education Network	AU
QGPOP	Kyushu Gigapop	JP
JGNII	Japanese R&E Network	JP, US, TH, SG

Table 4. Network Name

OSI Model			
	Data unit	Layer	Function
Host layers	Data	7. Application	Network Process to application
		6. Presentation	Data representation and encryption
		5. Session	Interhost communication
	Segment/Datagram	4. Transport	End-toend connections and reliability
Media layers	Packet	3. Network	Path determination and logical addressing
	Frame	2. Data Link	Physical addressing (MAC & LLC)
	Bit	1. Physical	Media signal and binary transmission

Table 5. Open Systems Interconnection (OSI) Basic Reference Model

10. References

Academia Sinica, (n.d.). The December 26, 2006, Pingtung, Taiwan earthquake, In : *Academia Sinica*, 26.07.2011, Available from :
http://www.earth.sinica.edu.tw/~smdmc/recent/2006/200612261226.htm

Academia Sinica, (n.d.). The December 26, 2006, Pingtung, Taiwan earthquake, In : *Academia Sinica*, 26.07.2011, Available from :
http://www.earth.sinica.edu.tw/~smdmc/recent/2006/200612261234.htm

Asian Internet Interconnection Initiative Project, (n.d.). Asian Internet Interconnection Initiative Project, In : *Asian Internet Interconnection Initiative Project*, 26.07.2011, Available from : http://www.ai3.net/

Asia-Pacific Advanced Network, (n.d.). Welcome to Asia-Pacific Advanced Network, In : *Asia-Pacific Advanced Network*, 26.07.2011, Available from: http://www.apan.net/

Asia-Pacific Advanced Network Japan, (n.d.). APAN-JP Network Traffic Weathermap, In : *APAN Japan Consortium*, 26.07.2011, Available from :
http://monitor.jp.apan.net/weathermap/

CISCO, (n.d.). Cisco IOS NetFlow In : *Cisco*, 26.07.2011, Available from :
http://www.cisco.com/en/US/products/ps6601/products_ios_protocol_group_home.html

Delivery of Advanced Network Technology to Europe, (n.d.). TEIN2, In : DANTE, 26.07.2011, Available from: http://www.tein2.net/

Fumagalli, A. & Valcarenghi, L. (2000). IP Restoration vs. WDM Protection: Is There an Optimal Choice?, *IEEE Network Magazine,* Vol. 14, No. 6, (December 2000), pp. 34-41, ISSN 08908044

Gerstel, O. & Ramaswami, R. (2000), Optical Layer Survivability: a Services Perspective, *IEEE Communications Magazine,* Vol. 38. No. 3, (March 2000), pp. 104-113, ISSN 01636804

GOOGLE, (n.d.). iperf – TCP and UDP bandwidth performance measurement tool, In : *Google Project Hosting*, 26.07.2011., Available from :
http://code.google.com/p/iperf/

Hanks, T. & Kanamori, H. (1979). A Moment Magnitude Scale, *Journal of Geogphysical Research*, Vo. 84, No. B5, (May 1979), pp. 2348-2350, ISSN 0148-0227

Iannaccone, G. ; Chuah, C. ; Mortier, R. ; Bhattacharyya, S. & Diot, C. (2002). Analysis of Link Failures in an Backbone, *ACM Internet Measurement Conference*, Marseille, 2002

International Organization for Standardization, (n.d.). ISO - Maintenance Agency for ISO 3166 country codes - English country names and code elements In : *International Organization for Standardizaation*, 26.07.2011, Available from : http://www.iso.org/iso/en/prods-services/iso3166ma/02iso-3166-code-lists/list-en1.html

INTERNET2, (n.d.). Bandwidth Test Controller (BWCTL), In : *Internet2*, 26.07.2011, Available from : http://e2epi.internet2.edu/bwctl

Konishi, K. (2007). Cables restoration, *23rd APAN Manila Meeting*, Manila, Feburary 2007

Kurokawa, Y. (2006). Registration of Policy Routing Choices , *22nd APAN Singapore Meeting* , Singapore, June 2006

Kurokawa, Y. (2010). Utilization of Compath Tool in Multiple Submarine Cables were cut off near Taiwan on August, 2009, *29th APAN Sydney Meeting*, Australia, February 2010

Kyushu GigaPOP Project, (n.d.). QGPOP (Kyushu GigaPOP Project), In: *QGPOP (Kyushu GigaPOP Project)*, 26.07.2011, Available from: http://www.qgpop.net

Lougheed, K. & Rekhter, Y. (1989), Border Gateway Protocol (BGP), In : *RFC 1105*, 26.07.2011, Available from : http://datatracker.ietf.org/doc/rfc1105/

Markopoulou, A. ; Iannacccone, G. ; Chuah, C. ; Mortier, R. ; Bhattacharyya, S. & and Diot, C. (2004). Characterization of Failures in an IP Backbone, *IEEE Infocom*, Hong Kong, 2004

Oetiker, T. (n.d.). MRTG - Tobi Oetiker's MRTG - The Multi Router Traffic Grapher, In : *MRTG*, 26.07.2011. Available from : http://oss.oetiker.ch/mrtg/

Oetiker, T. (n.d.), RRDtool - About RRDtool In : *RRDtool - About RRDtool*, 26.07.2011, Available from : http://oss.oetiker.ch/rrdtool

Ramamurthy, S. ; Sahasrabuddhe, L. & Mukherjee, B. (2003). Survival WDM Mesh Networks, *IEEE/OSA Journal of Lightwave Technology* , Vol. 21, No. 4, (April 2003), pp. 870-883, ISSN 0733-8724

Research and Education Network Operators Group, (n.d.) In : *RENOG*, 26.07.2011, Available from : http://www.renog.org

Robb, C. (2006). Global Research and Education Network Routing Issues, *2006 Spring Internet2 Member Meeting*, Arlington VA, April 2006

Saharabuddhe, L. ; Ramanurthy, S. & Mukherjee, B. (2002). Fault Management in IP-Over-WDM Networks: WDM Protection vs. IP Restoration, *IEEE Journal on Selected Areas in Communications*, Vol. 20, No. 1, (January 2002), pp. 21-33, ISSN 0733-8716

Sakiyama, R. ; Okamura, K. & Lee, Y. (2006). Visualization of Temporal Difference of BGP Routing Information, *APAN Network Research Workshop, Singapore*, Singapore, June 2006

Sharma, V. & Hellstrand, F. (month 2003). Framework for Multi Protocol Label Switching (MPLS)-based Recovery, In : *RFC 3469*, 26.07.2011, Available from: http://datatracker.ietf.org/doc/rfc3469/

Varadarajan, S. (1997). Virtual Local Area Networks , In: *Virtual Local Area Networks*, 26.07.2011., Available from : http://www.cs.wustl.edu/~jain/cis788-97/ftp/virtual_lans/index.htm

8

Simulating Collective Behavior in Natural Disaster Situations: A Multi-Agent Approach

Robson dos Santos França[1], Maria das Graças B. Marietto[1],
Margarethe Born Steinberger[1] and Nizam Omar[2]
[1]Universidade Federal do ABC
[2]Universidade Presbiteriana Mackenzie
Brazil

1. Introduction

The usage of simulations has been improved for quite some time. From mechanical artifacts that attempt to mimic a certain dynamic event using known physical properties up to complete representations of virtual worlds based on real life events which were augmented by concepts in order to prove a theory or to test a specific scenario. The key words here are "modeling", "constructing a simulacrum", "experimentation" and "evaluation". Simulations allow any researcher to explore, try out new ideas, check some theories in a controlled environment before testing in real life, and so forth. Psychology deals with individuals, Sociology with the study of human groups and the formation of institutions, both, individually, were not enough to study the humanŠs social behavior. All human sciences tried to create theories about reality, searching for well-defined and established patterns. The non-conformity with such patterns is considered a mistake, or even a wrongdoing. Taking a whole new approach, the field of Collective Behavior deals with human groups and collectivities that contradict or reinterpret societyŠs norms and standards. Crowd behavior has been studied by many researchers. Theoretical models have been established to understand them. This chapter will present a simulation model for panic in crowds phenomena based on the symbolic interactionism approach. Section 2 will present a review of the main concepts of Sociology and Collective behavior and establish a framework to be used in the model of crowd to be simulated. Section 3 will present a computation model and a simulation model of panic in crowd phenomenon, both in its theoretical aspects and its practical issues. The collective behavior studied in the previous section will be used as basis for the simulation model. Also, the main concepts regarding multi-agent based simulations will be presented. The model simulated have been applied to a fire incident and validated. Section 4 presents a generalization of the model proposed and delineates a future application for other kind of disasters as earthquakes. Section 5 shows some conclusions about the study here presented.

2. Sociology and collective behavior

Sociology deals with the study of human groups and the formation of institutions (dos Santos França, 2010; Merton, 1968). Its origin came from Comte, Spencer and other 19th century researchers' need for a distinct perspective of the human behavior that derived from the individualistic studies that had been performed previously. For instance, Comte stood out that the human mind could only develop in a social environment. Thus, following this premise, Psychology was not enough to study the human' social behavior (Turner & Killian, 1957).

At first, Sociology was focused on culturally-oriented groups or social groups which behavior follows established rules. Because of such interpretation, some spontaneous and unorthodox social actions were perceived as abnormal and unstable or as exceptions that did not draw further attention. Sociology, as a science, attempted to "frame" reality into well-defined and established patterns. If a certain social action could not fit into any of such patterns, the action was considered a mistake or even a wrongdoing until society accepts the new behavior and embraces it. Such acceptance could take decades or never happen.

Taking a whole new approach, the field of Collective Behavior deals with human groups and collectivities that contradict or reinterpret society's norms and standards. These collectivities' behavior is not entirely detached from the socially accepted behavior discussed earlier. However, collective behavior deals with social groups that deny or reinterpret society norms and standards. Ralph Turner and Lewis Killian at (Turner & Killian, 1957) defined collective behavior as *"the set of social behaviors which the usual conventions stop driving the actions and the individuals transcend, exceed or collectively subvert the standards and the institutionalized structures"* (dos Santos França et al., 2009). This definition implies that the individuals engaged in a collective behavior are no longer bound to the rules and norms of society and they are free to act the way they intended even if such behavior is not socially accepted. At first, their actions are related to the institutionalized and established actions found in Society. But, as soon their need for socially unaccepted actions is reached, they start to bend and to overrule the norms that were built by society, creating their own.

This sort of human group might happen due to many reasons, including by hazardous events, whether they are natural or human-induced. Also, their structure and formation follow a pattern that was mapped by some researchers. Finally, such mapped patterns could be used to understand disasters by a distinct perspective: how people react in a hazardous event and how this could be simulated in order to decrease material and human losses. The simulation model presented in Section 3 deeply applies the information described in this section.

2.1 Crowd simulation: Theoretical elements

The understanding of the panic in crowds' phenomenon relies on the study of the collective behavior phenomenon. Thus, a historical overview is presented in the following sections, along with modern studies about panic and disasters, especially how people behave under such conditions. The following subsections show a historical overview of some studies of the collective behavior field and the theories that will be employed in Section 3 to build the simulation model.

2.1.1 Historical abstract

The collective behavior was studied in distinct ways through the ages. Initially researchers such as Tarde and Durkheim developed social theories in order to justify the actions performed by offenders or as a mean of explaining how an isolated individual could have a socially accepted behavior and the very same individual could be able to participate in criminal acts when he is in a collectivity.

Emile Durkheim claimed that the group was important to understand the individual's behavior. Culture would be formed by the combination of personal minds instead of a chain of imitations from one subject by the other members of the group. This was one of the early conceptions of the group mind, a supra-personal entity which has an autonomous existence from the composing members of the group (Durkheim, 1895). In other words, the individuals engaged in a collective behavior unconsciously help to form the group mind that guides their actions.

Following an opposite direction, Gabriel Tarde considered that the social behaviors happen due to man's natural inclination to mimic others. For Tarde, the interactions among individuals worked only to spread the mimic's individual results and the interactions were not responsible by their formation. According to Tarde's approach, collective behavior describes the person's socially anomalous behavior into a group and collective context and in situations not induced by criminal activities, such as the tulip mania (Mackay & Baruch, 1932) or the great social movements, such as the fall of the Bastille (Tarde, 1890; Turner & Killian, 1957).

2.1.2 Collective behavior development

After a criminal approach for the collective behavior, some researchers analyzed the collective behavior phenomenon in an individualized and superficial way, such as Sigmund Freud (Freud, 1955). However, some other researchers such as William McDougall and Gustave Le Bon developed the collective behavior studies further by creating an early classification of the phenomena, as well as a detailed profile of each member of the collectivity, but also taking into consideration that the collectivity itself has its own specific features. This second attempt to understand the collective behavior phenomenon followed a psychological standpoint (dos Santos França, 2010).

Le Bon is considered one of the founders of the collective behavior studies and he was one of the firsts to use the term crowd to describe the collectivities, developing the Crowd Psychology and treating the crowd as the prototype of all group behaviors. The focus of his studies was the social behavior by using the "the crowd mind" theory. For Le Bon, the main features of the crowds were:

- The decreasing of the conscious personality along with the prominence of the unconscious one;
- The ideas and feelings of the members of the crowd are guided by suggestion and contagion;
- The trend to put suggested ideas into action.

A rough classification of the crowds was also proposed by Le Bon. Such classification was based on how the crowd was conceived and its main actions, and it can be summarized as follows:

Active crowd Crowds that act together with a strong sense of coordination. Examples include mutinies, lynching mobs and rebellions;

Casual crowd A crowd formed with no specific goal and coordination, acting at the same time and place for a short period. For instance, a crowd watching a display window being decorated;

Conventional crowd When a group of people gather themselves for a specific goal, sharing feelings that drive the actions of the whole group, such as what happens in an audience for a soccer game or any other recreational activity;

Expressive crowd A group of people gathered to move, make gestures together but for individual achievements, such as the dancing crowds at carnaval and some religious groups;

Panic crowd A panic crowd is formed when people are exposed to a dangerous situation and that leads them to create the perception of need to stay away from danger in a social and shared way, such as earthquakes and fires (dos Santos França, 2010; dos Santos França et al., 2009).

The psychological approach for the collective behavior emphasizes the lost of personality, the liability being empowered by the collectivity and the fact that such collectivity is guided by some kind of collective mind (similar to Durkheim's). Le Bon's vision also had the collective (or mass) psychology and the phenomenon of contagion in a primitive form (Le Bon, 1896).

The mass psychology was important for the development of the collective behavior studies because it was the first attempt to establish, classify and broaden such studies. However, the followers of this particular approach still treated the members of the crowds as society outcasts due to gender, race or civilization level. That implies that the only the civilized western individuals were considered truly civilized. Women, children, the mentally impaired and the individuals that belonged to a race other than white were marginalized and the mass psychology theories were used to justify and amplify such condition, as tools to "domesticate" and to "civilize" such groups, so they could act under the control of a leader such as Napoleon or Alexander, the Great (dos Santos França, 2010).

2.1.3 Symbolic interactionism and emergent norm theory

The criminal and psychological approaches for collective behavior used the abnormal, the unusual, the uncommon to establish a line, a threshold between the socially accepted behavior and groups (studied by Sociology) and the socially unaccepted behavior and the human groups that engaged in such behavior. Some researchers at the University of Chicago developed a distinct way to see and understand the collective behavior.

Robert Park and Ernest Burgess wrote a whole chapter about collective behavior in their book *Introduction to the Science of Sociology*. In that chapter, the concept of social contagion was described as an element to spread a cultural matter, being compared to the fashion phenomenon and inducing people's feelings. Thanks to Park and Burgess' work (and similar works released almost at the same time) collective behavior was related to social phenomena other than criminal activities and psychological issues. Also, the individual engaged in collective behavior could belong to any social group, according to certain social-cultural contexts (dos Santos França, 2010; Park, 1939).

Park also introduced the concept of "milling": a collective movement that represents fear or discomfort. The social unrest can amplify the fear which, in turn, leads the group to a tension state. Such unrest, even if it is merely mentioned, amplifies the fear. Thus, the milling and the social unrest make a vicious circle and their interaction becomes a circular reaction that increases the tension in the group and creates an expectation that mobilizes the group members for the collective act (Park, 1939).

Herbert Blumer was a student of Robert Park and carried on his research. George Herbert Mead was also Blumer's teacher and developed the social act, a noticeable external behavior.

With that theoretical basis, Blumer coined the Symbolic Interactionism, which society is built by the interaction among people that, when they are about to act, take into consideration the actions and features of the other individuals, a symbolic interaction driven by each individual meaning developed during the interaction process (Borgatta & Montgomery, 2000).

According to Blumer, Symbolic Interactionism is based on three premises:

1. The persons interact by the meaning of their world's objects (tangible, abstract or social), both individually and collectively;
2. The meaning of the objects is built from the interactions among individuals;
3. During the interaction, individuals use an interpretative process to change such meanings.

The Emergent Norm theory was proposed by Lewis Killian and Ralph Turner and it was presented in (Turner & Killian, 1957). Based on Blumer's Symbolic Interactionism, it also considered that the collective behavior was the outcome of the interactions among persons able to assess the received information which leads to an interactive cognition. This approach analyzes the agents' features that aided in the formation of the social systems in a micro level, as well as the behavioral patterns in a group level.

Therefore, the emergent norm approach deals with the formation of the collective behavior by the micro level interactions of the collectivity members and the advent of patterns and norms triggered by these interactions. Although there is no emphasis in the definition of the social systems (as seen in (Luhmann, 1996)), the interactions and the complex behaviors formed by them allow the collective behavior to be seen as a complex system because from its micro-level interactions - simple by nature - complex behavioral patterns emerge, and such patterns cannot be noticed by just analyzing each individual alone (dos Santos França, 2010).

2.1.4 Other approaches for collective behavior

Due to the need of creating a symbol and meaning system, Blumer' symbolic interactionism has some unclear basic points related to social interaction:

1. How individuals relate to each other in spite of their differences;
2. How the social relation comes from the orientation to the other in each attendant (Vanderstraeten, 2002).

These points were addressed by Talcott Parsons in his studies about social groups, which led to the Structuralism Approach for the collective behavior phenomenon.

The Structuralism Approach turns over the concept described in the previous section by highlighting the social structures' studies and their impact on the individuals. The focus lies on the social structures that triggered the phenomenon and the structures affected by the members of the collectivity, using the macro level elements to think about the micro level elements and behavior. Therefore, the social structure is analyzed as deep as possible. Any behavior that subverts the established social order is reviewed by observing how the social structure and the collectivity respond to that (dos Santos França, 2010).

Neil Smelser was a researcher at the Oxford University, and he was Talcott Parsons' student at the time. Enhancing Parsons' collective behavior studies (Parsons, 1937), Smelser pointed out that, although rumors, panic or lunatic conditions, commotion and revolution are unexpected and surprising, they happen regularly (Smelser, 1963). He also stated that as much institutionalized the behavior is, it will become less distinguishable in a social point of view. The purpose of collective behavior, according to Smelser, is the resettlement of the social order that was shook by a tension on the elements that make the social structure. The resettlement induces people to act in a collective and rational way. After that, social norms and institutions are crystallized due to the comeback of the social order or by the formation of a new one. This shows Smelser's top-down approach for the collective behavior phenomenon (Smelser, 1963).

2.1.5 Panic in crowds

Panic in crowds can be triggered by various factors, such as natural threats (floods, earthquakes, volcanic eruptions), threats induced by man (terrorist attacks, lost of the social control by State), among others. In a panic situation there is always an imminent risk and the urge to act by the individuals (dos Santos França et al., 2009).

Killian and Turner also studied the behavior of individuals during crisis. In (Turner & Killian, 1957) the micro interactions are the key elements for the changes in the society. The same would happen with culture that changes thanks to each person, even if that happens in an unusual and unconscious way. According to Killian and Turner, it is in the reaction of the individuals in critical and unstructured situations that the basis of the collective behavior can be found. Such personal responses should be accepted as a required background for the study of the development of new norms and social structures.

Three kinds of individual reactions were found by Killian and Turner. The first kind of reaction is Defense: people act in a limited fashion, unable to comprehend what happened and to deal with new situations, and some of them will be in shock, even with no physical damage. On the other hand, there will be others that become more suggestible and readily accept commands from somebody else (Turner & Killian, 1957).

The second kind of reaction that usually happens after the shock from a violent accident is an impulsive and apparently irrational action. The individual acts apart from the environment and the other individuals, with actions entirely out of his normal self, in some kind of "super focus". Even though that individual is aware of what happens in the environment, his actions are directed towards a specific spot inside the event, acting in a conscious way. It seems that in such behavior there is an attention strain. Thus, the individual does not think about the consequences of his acts in the same degree of his actions in ordinary conditions (Turner & Killian, 1957).

The final kind of individual reaction found by Killian and Turner is the fear. A critical situation is known to pose as a threat to the individual's life or values. Thus, fear is the most common reaction in panic situations, even if such situation is not real.

Fear can be shown in many ways, from internal changes in the emotional and psychological state up to despair, whimper and foray, and it increases whenever the danger is unknown. Uncertainty leads to insecurity since the person does not have enough information to take the right decision in the new context. A person is less afraid of a dangerous situation than the lack of information of the present condition and its uncertainty (Turner & Killian, 1957).

Panic in Crowds phenomenon has been studied by many researchers, mostly to understand its inner workings and specially to prevent the dangerous events to start it or to alleviate its effects if it is unpredictable. Enrico Quarantelli is a researcher that provided some essays about disasters and panic in crowds' phenomena.

In (Quarantelli, 1975) Quarantelli identified a certain set of prejudgments related to how people observe the crowds' behavior in panic situations:

- People would behave "irrationally", out of control;
- Thanks to media and films, panic is associated with despair, paralysis (shock) and an instinctive behavior caused by the panic itself, forcing a subtle mind changing similar to the one found in "Strange Case of Dr. Jekyll and Mr. Hyde" by Robert Louis Stevenson.

These prejudgments are passed to the safety and damage control personnel, such as firefighters, police officers, public managers, among others. For example, the fear of inducing panic just by informing people about the hazardous event could be more dangerous than the life-threatening event itself. Even with relevant and crucial information for crowd control and to minimize material and human losses, the fear of generate more panic could block the right actions at the appropriate time, which in a panic situation could be disastrous. For Quarantelli, the mere mention of a dangerous situation does not trigger or amplify the crowd's panic state (Quarantelli, 1975).

In spite of what was proposed by the early researchers of collective behavior such as Gustave Le Bon, the human behavior during crisis is controlled instead of impulsive, it uses the right means to achieve its goals and it is organized and functional most of the time. However, that does not mean that an irrational behavior is avoided during the crisis; the incidence of such behavior is lower than what was intuitively observed.

Just like the other collective behavior and panic researchers, Quarantelli also provided the panic's main features, based on his studies and the analysis of other studies from Japan, France and England, and they are the following:

- A person in a panic in crowds' situation deals with fear instead of anxiety;
- The future is more important in such situations than the past;
- There is a trend to focus in a specific dangerous spot instead of a general threat;

- The members of the collectivity define the situation as dangerous and identify a direct threat for their survival (Quarantelli, 1975).

Quarantelli stressed that individuals keep their rationality and sociability during their escape from hazardous places: they avoid obstacles and other people as much as possible. The individual still can force his way over the others, but that will happen only in extreme conditions (Quarantelli, 1975).

This chapter will present a simulation model for panic in crowds phenomena based on the symbolic interactionism approach. The panic phenomenon works as follows.

Initially, people are in an **ordinary condition**. In that condition, social structures and norms are lined up to what is accepted by society. At the moment disarray in the established social structure is noticed, individuals start feeling uneasy and apprehensive, trying to understand the ambiguous situation that occurred. A disturb is an event that shows itself as an imminent threat to the individual's life, such as a fire alarm, a smoke cloud or objects falling from the shelves, and such event calls up the person and compels him to act, leading to a **social unrest**.

After that, the persons search information that could help them in redefining the present context. They become more likely to rumors because of the feeling of uncertain and insecurity. The conventional behavior starts to break down. The need to comprehend the situation increases, so they engage in a **milling** process, watching the other individuals' reactions and comparing those reactions with your own set of expectations. Also, a need for a sanctioned and socially-built meaning arises into a relatively non-structured situation (Turner & Killian, 1957). Milling is substantial since it makes the individual focused to the situation and the actions performed by the collectivity, removing the focus out of him. Due to the fact that the focus now lies on them, the individuals reply faster and directly to each other, setting up the environment for the shaping of a shared knowledge of what is happening. From that point, the collective enters the **collective excitement** stage, when the group blends and synthesizes the personal representations, helping in the formation of a collective representation/image of the situation. The individual's susceptibility is enhanced by this shared representation, which also decreases his capability of making distinct impressions from the collectivity.

Thus, the individual could follow a socially forbidden line of conduct that he could not conceive and perform, such as pushing and running over people. **Social contagion** starts as an intense form of collective excitement, it starts fomenting a fast propagation of the collectively formed representation, strengthens the social cohesion and prepares the crowd for a collective action. Finally, after a collective representation of the situation is built by the individuals, it is possible to pick an action and execute it. Up to this moment, the collective crisis started by a struggle for survival comes to an apex, and the **collective panic** is installed. Considering that the crowd members do not share conventional expectations about how they are supposed to behave, the outcomes are uncertain. Figure 1 shows an overview of these stages.

2.2 Multi-agent based simulation: Usage and features

A simulation is the representation of a contextualized system into another context. This description applies to any kind of simulation, not just computer simulations. The Apollo space mission had applied simulations to evaluate techniques and devices before the real

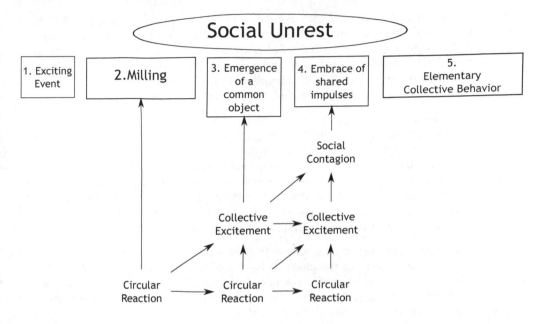

Fig. 1. Blumer's Collective Behavior Stages (McPhail, 1989).

mission was performed. Since the mission posed a great risk and there was no much room for on-the-fly modifications, everything must be tested and checked beforehand.

In Ruas et al. (2011) simulation (especially multi-agent based simulation) is regarded as a third way of doing science. While induction studies the whole by a sample, deduction does the opposite. Simulations get the best of induction and deduction at the same time: the general and macro-level of a process provides the framework, while the interactions among simulation elements show the micro-to-macro transition and the emergence of behavioral patterns, as in the induction process.

Throughout this section, the simulation process is described. The focus will be on the multi-agent based simulation, which will be applied in the model presented in the next section.

2.2.1 How a simulation is designed

The design of a simulation is the building of a model that will be able to mimic the operational and dynamic features of a real system. This model allows a deeper study of the system in a controlled and isolated context Zeigler et al. (2000). This usually poses as a requirement for some systems since the analysis and observation of certain phenomena and their activities can be impossible, impractical or hazardous.

There are two major approaches for computer simulations. The first approach uses differential equations and other mathematical formulas to build the simulation model. The simulation execution becomes the evaluation of such formulas and the iterative resolution of the differential equations. Such approach is named analytic and it has its value and it is quite practical and useful for certain applications. However, it usually lacks a detailed vision of what happened, working as a "black box". For some simulations, this is not an issue because the only thing that matters is the final result and not the mid-steps required to achieve it. Also, this sort of simulation usually deals with a continuous stream of time. Since there is no need to observe the simulation's inner steps, a continuous approach is more logical.

On the other hand, a second approach for simulations uses a set of autonomous modules (programs) called agents. The resemblance of agent based technologies and a realistic social system model has created a new scientific field with a strong emphasis on the interdisciplinary called Multi-Agent Based Simulation (MABS) Cohen & Felson (1979). It is a collective effort to integrate scientific areas and the usage of computational technologies that were previously applied to other tasks, such as networking. The main purpose of MABS researchers is to create and study computational models for simulation taking the technical and theoretical infrastructure of the Distributed Artificial Intelligence into consideration.

Based on such approach, the simulation model represents a specific target system that allows (i)the observation and study of the global behavior of the modeled system under certain criteria and (ii) the analysis of the consequences of the changes in the system's internal components Gilbert & Terna (2000), which implies that MABS can be used to detect emergent patterns and how changes interfere on the agents' behavior. Ruas et al. (2011).

2.2.2 Agent and multi-agent based simulations

A specific definition of Agent describes it as a discreet entity with its own goals and behaviors, and also internal states and behavior rules that allow the interaction with the other agents and with the environment. Another definition can be found in Russell & Norvig (2004), and states that "*An agent is anything able to perceive the environment through sensors and to act upon the environment by actuators*". Once more, the emphasis lies on the agent, the environment and the relationship between them. Whatever entity that needs to be considered in the simulation by its autonomy, by its independence in the decision making and by its ability to interact in the environment can be seen as a simulation agent dos Santos França (2010).

Agents must have autonomous actions, and such actions must happen synchronously with an event-based time scheduler, that will serve as an observer and a time and step manager along with the agents.

The main concept behind a Multi-agent simulation model is to simulate an artificial world which is made of computational interactive entities. Simulation is then created by the transposition of entities (or sets of entities) and the interaction among such entities from the target system to the artificial world Dimitrov & Eriksen (2006).

The multi-agent based simulations have an adequate infrastructure for modeling, studying and understanding the process related to complex social interactions such as coordination, collaboration, group formation, conflict solving, among others. Thanks to the relationship between local and global behaviors and the analysis of the agents' influence over themselves and the environment, it is possible to analyze the social interactions, which leads to cause-effect relations of how agents' components affect their behavior, how such behavior affect the group and, likewise, how the group itself affect these components. The analysis of the situation implies the analysis of the environment where the agents are located, the decisions taken by those agents, how such decisions affect the environment and the other agents and how the groups of agents can affect the agents' internal attributes dos Santos França (2010).

The multi-agent model for the panic in crowds phenomenon described in Section 3 belongs to the social-cognitive model class David et al. (2004) because such models have their focus on formalization and testing of theories, models and hypothesis related to theoretical-structural aspects of social systems. The main concern in this class of simulation models is the dynamic behavior of the simulation instead of an exact and perfect outcome analysis. For this class, the straight comparison of the simulation outcome and some empirical data could render pointless because the target system cannot be fully represented in any form, especially if the system is complex. Therefore, the subject of study of the panic model described in Section 3 matches the structural logic of the target system and it works in two dimensions:

1. To propose new structures or replacements for social systems, checking their viability and working;
2. To get a better understanding of the social, psychological and anthropological bases which sustain and direct the panic collective behavior dos Santos França (2010); dos Santos França et al. (2009).

2.2.3 Conceptual model

The multi-agent based simulation models share some common features. The model has autonomous and heterogeneous, they are not under a central authority's orders because they are built to be self-organized and with local interaction rules.

The agents are in an environment that encourages the interaction among agents so that the model can fulfill its main goal: to be open to the emergency of phenomena due to the interaction among agents and the environment, which makes the multi-agent based simulations work as complex systems. A system is said to be "complex" if its overall behavior cannot be described by just looking at its inner elements' behaviors. In order to understand a complex system's behavior, the observation of the emerging patterns created by the agents' interactions is required.

The following list has some situations which the agent-based models are more suitable for watching the emergent behavior da Silva et al. (2008):

1. When there is a substantial need to design heterogeneous agents populations, and such heterogeneity enables the modeling of agents with rationality and clear and distinct behaviors;
2. Every time the agents' interactions are discontinuous, non-linear such as the individuals' complex behavior, which make the process harder for classical analytic ways;
3. Whether the agents' interactions' topology presents itself as heterogeneous and complex, such as the social processes, in specific the inherent complexity of the physical and social networks.

2.2.4 Computational model

The Computation Model is the representation of the Conceptual Model in a programming language or simulation tool so that model can be evaluated and analyzed. The process of building the computation model is similar to application software development. Usually, the same tools are employed, such as text editors, integrated development environments (IDE's), testing tools, graphics libraries and so on.

The usage of a simulation framework allows the developer to keep her focus on the model and the simulation details instead of the programming language and running environment details. There are many simulation frameworks available, such as Repast, NetLogo and Swarm. Most of these frameworks can be combined with other tools and libraries. For instance, the Swarm Framework SwarmTeam (2008) is written in Objective-C and it also supports Java for simulation building. Since a simulation could be written in Java, it would be possible to use Java-based libraries - such as JESS Friedman-Hill (2009) - to enhance Swarm agents and the simulation as a whole.

Usually, the simulation developer must create objects that represent her agents. The agents' variables become the objects' fields. Likewise, the agents' actions become methods.

The simulation developer may face some challenges, such as:

- The choice of a random number generator or the creation of a customized one. Some frameworks provide a generator. However, for some specific situations, a generator created from scratch must be required. Although they are called "random", in reality they are pseudo-random, and that happens for a reason: a simulation (even multi-agent based) usually requires numbers that set the simulation up and could be fed during the simulation process. The developer must be in control of the numbers' generator to avoid an excessively predictable behavior and a fully random behavior;
- The usage of supplementary tools that might aid the simulation process and the post-process. These tools include databases, graphical viewers and network facilities, among others. Just like the random number generator, some frameworks provide these tools. It is up to the developer to choose either the tools found in the framework or to create them on her own , or even mix the best tools from both sides;

2.2.5 Verification and validation

An aspect of great relevance in simulations is how accurately the conceptual model and the computational model depict the target system. Two processes can be used to check such confidence.

The first process is called validation. Its main purpose is to make sure that the conceptual model represents the target system in a certain (and desirable) level of precision and to show whether the simulation's results match the target system Ruas et al. (2011).

Verification goal is to certify if the conceptual model was rightfully implemented (translated) in the computation environment. Since a computer simulation works as a software application, it is possible to use software engineering tools, such as unit tests, to certify that the behavior designed for the simulation (found in the conceptual model) really happens in the software execution.

The validation process aims to certify that the conceptual model represents the target system in an acceptable degree of adherence. Thus, the validation processes fundamentally addresses a specific question: Does the simulation outcomes correspond to those from the target system? On the other hand, the verification process' main purpose is to assure that the conceptual model was correctly translated to the computational environment. Specifically, a multi-agent simulation model is based on the concept that it is feasible to simulate an artificial world inhabited by interactive computational entities. Such simulation can be achieved by transposing the population from a target system to its artificial counterpart. In that sense, an agent is similar to an entity or a group of entities of the target system. Moreover, agents can be of distinct natures and granularities, such as human beings, robots, computer algorithms, inanimate objects and organizations. (Ruas et al., 2011)

3. A simulation model for panic in crowds phenomenon

3.1 From theory to practice: Conceptual model

In order to build a conceptual model for the panic in crowds' phenomenon the following elements will be discussed:

1. The architecture of the agent that represents a person in a panic situation;
2. Three environments (General, Physical, Communication) where the interactions' main aspects happen;
3. A socially built system - Collective Mind - that describes how individual representations are transformed and synthesized by the group so they form a shared context (dos Santos França et al., 2009).

This model proposes the interactionism approach presented by authors such as Blumer (Section 2.1.3). A generalized flow based on that theory is shown in Figure 2. It is worth noticing that the exhibition of the steps is in a sequential order for didactical purposes. However, it is possible that a person follows a distinct order, not performing some steps or repeating others.

3.1.1 Model's environments

3.1.1.1 General environment

This element represents a general overview of the environment where all the interactions among agents will happen, and it has the Physical Environment, the Communication Environment and the Collective Mind. Its purpose is defining the boundaries of the other environments and their linking points. Figure 3 shows the proposed diagram for the relationship of these elements.

3.1.1.2 Physical environment

The Physical Environment describes the space where the physical interactions among agents occur, as well as the interactions between the agents and the other objects such as obstacles and walls. There are specific spots for the threat and the exits. Figure 4 shows this environment

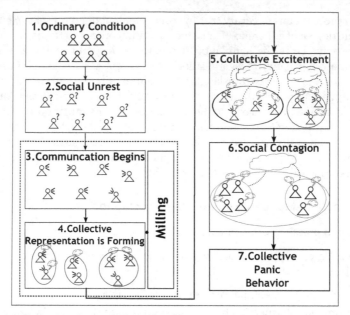

Fig. 2. Collective Behavior General Flux (dos Santos França, 2010).

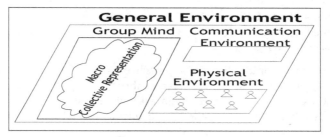

Fig. 3. General Environment and its Components (dos Santos França, 2010).

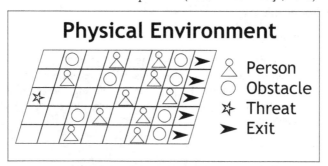

Fig. 4. Physical Environment (dos Santos França, 2010).

The agents can move in four directions (north, south, east or west). Besides, there is a chance of lane change according to agents' traffic during the simulation, which makes the agent moves diagonally if required.

This environment, along with the Person agent, also has the Obstacle, Threat, Exit, Milestones and Fire Spot agents. An Obstacle blocks people's passage, forcing them to dodge. In the model described in (Helbing et al., 2002), the building structures (walls and pillars) and the wounded and immobilized individuals are treated as obstacles.

The Threat agent is the element that triggers the exciting event in a panic situation. For the model described in (dos Santos França, 2010) in particular, it is a fire incident modeled by a structure that represents the environment's heat as a 2D grid. Such structure is responsible for heat diffusion between cells.

The Exit is the physical environment's safe haven. When the Person agent arrives on that place, he does not feel threatened and he gets disengaged from the collective behavior, which makes him no longer relevant for the simulation.

Milestones bound Threat's influence zones and they serve as reference to the emergent behavior analysis. Fire Spots are fire's control points that establish how far the fire went through the environment. Along with the Milestones, the Spots can help in outlining potentially safe or dangerous zones, working as if buzzers and visual alarms were triggered by a smoke detector. However they do not exist physically; they are just the representation of the agents' response to such elements.

3.1.1.3 Communication environment

The Communication Environment manages and serves as medium for the three communication forms among the agents: through the environment (physical perception), directly (sender/receiver model) and indirectly (dissipation/perturbation). In this third form, whenever an agent wants to communicate with another, it places the message on the environment (dissipate) and if another agent may be disturbed by that message or not. That occurs because it is not possible to control the expectations and actions from the other agents and assures that the communication will happen *a priori* Figure 5 shows the Communication Environment.

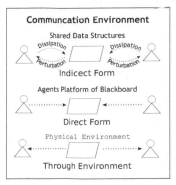

Fig. 5. Communication Environment.

A Blackboard system (Rich, 1988) was used for the direct and indirect messages. Physical information was gathered directly from the environment.

3.1.1.4 Collective mind

The Collective Mind manages expectation networks that are socially built. These expectations arise because the agents look forward to certain behaviors from other agents, as well as the knowledge found in the agents themselves that their actions can also be part of the expectations of the other agents. The Collective Mind also makes abstractions, generalizations and schemes from the individual expectations, taking control of the emergent process of a current context shared representation (dos Santos França, 2010).

3.1.2 Person agent architecture

The Person agent portrays an individual that will have a behavior related to the collective panic situation. Such behavior is directed by the Symbolic Interactionism and Norm Emergent theories described in Section 2.1.3. An overview of the agent's architecture can be found in Figure 6.

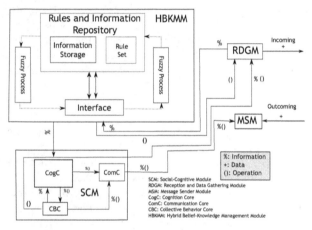

Fig. 6. General Overview of the Person Agent.

The Person agent is made of four modules: Reception and Data Gathering Module, Hybrid Belief-Knowledge Management Module, Social-Cognitive Module and Message Sender Module.

3.1.2.1 Reception and Data Gathering Module

The main function of the Reception and Data Gathering Module (RDGM) is listening to the environment, gathering relevant information, establishing the nature of such information and storing it in the Information Storage. This module has two cores: the Data Selector that scans data for the creation of a current situation portrayal, and the Information Analyzer that checks the information integrity in terms of syntax and semantics and passes the information to the correct information base: Knowledge Base of Belief Base. Physical information is always treated as knowledge and information provided by other agents is accepted as a belief until it is confirmed. The Social-Cognitive Module can also request data gathering on the environment if the agent needs some information from the other agents (dos Santos França, 2010).

3.1.2.2 Hybrid Belief-Knowledge Management Module

The module that manages the information bases and the rule set of the Person agent is the Hybrid Belief-Knowledge Management Module (HBKMM). The information can be modeled by a stochastic, logical or a fuzzy approach that is used because some information kept by the HBKMM is imprecise and incomplete by nature. Since there is fuzzy information, a Fuzzy Process is also required so the information can be fuzzified and de-fuzzified according to the agent's demand (dos Santos França, 2010).

The Rules and Information Repository groups the Information Store that keeps the beliefs, knowledge and the micro collective representation, and the Rule Set which holds the Person agent's general behavioral rules.

The Information Storage holds deterministic (analyzed using equations or algorithms), probabilistic (that follows a stochastic uncertainty that defines whether it is truth or not - or fuzzy information - that deals with the possibility of the information being truth or not in a scale.

The Knowledge Base stores the information that are treated as secure and confirmed by the agent. In model shown in (dos Santos França, 2010), if the information requires physical evidences, but the agent could not be able to get the evidences using its own perception, then the information is treated as a belief. Thus, the agent's variables can do a status change (from belief to knowledge) during the simulation.

The agent' personal features define the state of the agent. They are variables that change their value according to the information gathered by the agent and the agent's actions and processing. There are also constant features that were defined before the simulation started and their values do not change during the simulation.

An example of the agent's variable is the Dangerousness. It is a complex variable that relies on other variables of the agent, such as distance from the threat, health, the agent's experience on this kind of hazardous phenomenon, among others. Considering that this variable has fuzzy information, in order to be fuzzified and de-fuzzified a fuzzification table (Table 1 is used. Figure 7 shows how a graphical view of such table.

Zone	Value
0.00 ⊢ 0.20	Safe
0.10 ⊢ 0.50	Slightly dangerous
0.45 ⊢ 0.80	Mildly dangerous
0.75 ⊢ 1.00	Imminent Life Threat

Table 1. Dangerousness Levels for Fuzzification.

Fig. 7. Fuzzification Graph

The Rule Set has all the rules that the Person agent may perform during the simulation. An example of rule is "Establish the Agents' Pressure Limit" that updates how much pressure the agents can hold based on their individual size. Listing 1 describes how this rule is performed.

```
1  on the simulation's setup process
2  do
3          foreach agent in worldAgents do
4              agent.pressLimit = agent.size * 2 * PI * PRESS_LIMIT_FACTOR
5          endfor
6  end
```

Listing 1. Establish the Agents' Pressure Limit.

3.1.2.3 Social-Cognitive Module

This module is responsible for coordinating the agent PERSON other modules' actions, managing their autonomous and private process. It is made of the following cores: COGNITIVE CORE (CogC), COLLECTIVE BEHAVIOR CORE (CBC) and COMMUNICATION CORE (ComC).

The CogC stands in continuous processing, managing information and guiding actions so the agent can pursue its goals. As long as the individual is in a situation that does not pose as a threat to its life (see Fig 2, item 1), the CogC leads the agent to a certain behavior that it accepts the rules and roles established in the society. However, if an event that poses a threat is triggered, the CogC passes his duties to the CBC. This replacement makes the agent act in a collective way, engaging in the collective behavior. Also, the CBC deals with the agent's collective behavior state machine.

In order to quantify the threat, the agent checks his experience and the hazardous level he assigned for the current situation. Up to that moment, the functional rules remain strong, and the reactive ones still remain weak. The individual does not have enough information to go to a specific line of action. Thus, in order to go to the next step (social unrest), the uncertainty level assigned for the situation must be higher than a certain threshold, which implies that the agent doesn't know what is happening, so he feels that he needs more information about the event (dos Santos França, 2010; dos Santos França et al., 2009).

When the agent goes to the social unrest state (Fig. 2, item 2), he looks for information that helps him to analyze what is going on. Its uncertainty level rises since it is unable to understand the event by himself. Thus, he , so it engages in the milling process (Fig. 2, item 3). At this point, the agent increases his communication with the others, trying to build his own MICRO COLLECTIVE REPRESENTATION (Fig. 2, item 4). At the same time the personal value variable is affected, increasing the agent's acceptance for external thoughts. The agents become less aware of themselves as individuals and more aware of the others. The dynamic rules (e.g. learning how to perform an operational task) become weaker because the sense of urgency is stronger in a dangerous situation than in an ordinary condition dos Santos França (2010); dos Santos França et al. (2009).

Collective excitement (Fig. 2, item 5) begins when the permissiveness starts to interfere on the agent's choices. At this point the agents can choose socially unacceptable actions, such as running over people. Functional rules lose their strength (mostly because permissiveness is rising) and reactive rules get stronger.

When the agents define a goal and an object for action, the macro collective representation starts to develop and to establish.

This step is called social contagion (Fig. 2, item 6) because the communication and interaction among agents are in such condition that some individuals - not yet engaged in collective behavior - are attracted by the group, and they are induced to be part of this process. The reactive rules become the strongest rules for the agent. Since the permissiveness is high, the agents can choose actions treated as socially improper. Dynamic rules, such as learning how to escape are limited (dos Santos França, 2010; dos Santos França et al., 2009).

Finally, the collective panic behavior (Fig. 2, item 7) is installed when the agents choose a line of action to be followed by the collectivity. The agents are fully engaged in the collective behavior, and they will stay on that condition until they do not feel threatened.

The ComC receives all requests for communication from the CogC and the CBC and puts those requests in a queue for being dispatched by the MESSAGE SENDER MODULE.

3.1.2.4 Message Sender Module (MSM)

Whenever the agent needs to send a message to the other agents, this module is requested. The MSM receives the message from the COMMUNICATION CORE. Inside the MSM the MESSAGE FORMATTER prepares the message to be dissipated on the environment by encoding, adding other relevant data, such as the message format (using an ACL) and how it should be expressed in the environment: if it is a gesture or a speech and how the message mood is (lovely, cold, etc.) dos Santos França (2010).

3.2 Bring the concept to life: Computational model

The computation model is the transposition of the conceptual model to the computational realm. In order to achieve such transition, there are two major choices. The first choice is building the whole simulation program and framework by hand. In other words, the developer could write all the elements of the simulation and a framework to manage the simulation.

3.2.1 Implementation details

This simulation was entirely written in the Java programming language. As it was described in Section 2.2, each agent (Person, Exit, Threat and Obstacle) was modeled as a Java class.

The framework used to implement this model was the Swarm Framework, found at (SwarmTeam, 2008). The database engine used to store the simulation statistical data was the HSQLDB (Hsqldb Development Group, 2009), a free and open-source database engine written in Java.

A simple log system was also designed and it could be set up to store step-by-step state data for all agents or just for a set of them. The log data was stored using the YAML standard for better human readability than CSV or XML.

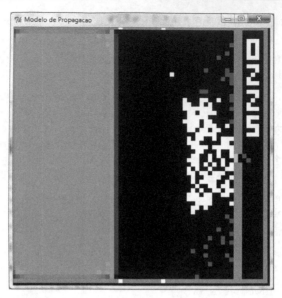

Fig. 8. Simulation Screen Shot

The Swarm Framework does not provide an Expert Systems' support, so the JESS (Friedman-Hill, 2009) (Java Expert Systems Shell) library was applied. FuzzyJ (Brown, 2009) was used for the fuzzy logic rules.

In order to keep the simulation "random" and controlled at the same time, a set of ten random seeds were chosen. Since the simulation was run ten times, for each simulation run a specific random seed was used to keep the simulation analysis consistent.

The usage of a multi-agent simulation framework as Swarm allows the developer to think more about the simulation itself rather than the crosscut concerns, such as graphics. Figure 8 shows a screen shot of the simulation. All the graphical elements were drawn by Swarm Framework. Each colorful dot represents an agent, while the red area on the left is the threat (fire in this example).

Since the model is social-cognitive, the best validation approach is by analyzing the dynamic behavior of the simulation and checking if such behavior is coherent with the theory. The data gathered during the simulation combined with its dynamic behavior is used to validate the conceptual model. Swarm displays the physical environment as an animated 2D grid (lattice), and such animation provides the dynamic aspects of the simulation.

4. Earthquake simulation model: A proposal for future works

The previous sections described the panic in crowds' phenomenon, both in its theoretical aspects and its practical issues. The collective behavior studies shown earlier were used as basis for the simulation model proposed in Section 3. Also, the main concepts regarding multi-agent based simulations were also presented. The computational model was tuned for a fire incident. Could it be feasible to do the same thing for earthquakes?

In order to answer this question, a discussion about the definition of disaster must happen. Once again, Quarantelli provided a study about disasters and earthquakes in (Quarantelli, 1981). The first part of the aforementioned paper pondered about the definition of disaster and how researchers usually face the matter.

According to Quarantelli, some researchers have a biased and habitual view of disasters which partially blinds them from other possibilities. There would be two ways of analyzing disasters: focusing in the agents that caused the disaster or taking a more generic approach.

Quarantelli identified seven conceptions of disaster. Each conception analyzes disaster events through different approaches. Some of these approaches are related, but they are focused in distinct elements of the disaster:

Physical Agents This conception accepts a disaster whenever its primary cause is identified. And it seems natural that the cause for an earthquake is different from the cause of fire. The focus is pointed at the physical agent that caused the disaster. Such agent (the cause) must be described in detail, and the knowledge about one agent does not help in analyzing another one. Distinct agents require completely distinct studies;

Physical Impact of the Physical Agent Whenever there is a noticeable physical impact in some part of the environment, the disaster is identified. The physical agent is no longer relevant, but how this agent affects the environment. Instead, how the physical agent's features in the geological, biological and social-technical spheres of the environment affect the impact becomes more relevant than the agent itself;

Assessment of Physical Impacts While the first conception deals with the physical cause alone and the second conception analyzes the impact of such agent in the environment, this conception understands the disaster by the assessment performed on the physical impact. Thus, an event can only be called a disaster if the physical impact crosses a certain benchmark or threshold defined in an assessment. For instance, an earthquake could only be called a disaster if its strength - measured in the Mercalli and Richter scales - goes beyond an established level and it becomes notable;

Social Disruptions Caused by an Event with Physical Impact For this disaster conception, if the physical impact also causes a social disruption of the social life - represented by dead bodies and wrecked buildings, for example - the event is treated as a disaster. Following this conception, in order to identify a disaster, a social disruption (disorganization) must happen due to some physical impact, and the disaster will be graded by the social disruption;

Social Construction of Reality in Perceived Crisis The previous concepts take the physical element into consideration for defining a disaster. It is assumed that some physical event happened and that triggered the disaster, be it directly, by its impact, by an assessment or by the resulting social disruption caused by the impact. The physical component takes distinct roles in each definition, but it must always be present. The conception of disaster as a social construction of reality takes the people's perception as the key element to identify some event as a disaster. There is no need for physical evidence. If people believe that the situation is dangerous and poses as a threat to life, property, well-being or social order, the event is accepted as a disaster. Quarantelli stated that this approach makes the disaster a relativistic term rather than an absolutist one. Different groups may interpret the same event as a disaster or not;

Political Definition Being slightly similar to the previous conception, the political definition claims that the disaster definition comes from a political standpoint, even if the event could be accepted as a disaster for the other conceptions. On the other hand, by political demand, a situation that could not be portrayed as disaster may be addressed as such. Quarantelli stated that for those who define disasters by this definition "the formal designation can make a difference in everything from mitigation and prevention, to response and recovery activities."(Quarantelli, 1981). Therefore, a political decision on the matter of disasters can make all the difference between prevention, fast response / recovery and further damage control;

Unbalance in the Demand-Capacity This final conception takes a disaster as a type of crisis situation or a social occasion. An event is considered a disaster if the demands for urgent actions due to a threat to high priority values and the resources available do not meet such demands. Quarantelli recalls Erwin Goffman when he used the term occasion, which is related to "non-routine and emergent collective behavior". Thus, if the situation requires an unusual and new social behavior to balance the needs and the resources found in the occasion, that situation leads to a disaster.

These concepts ranged from a purely physical approach to social related approaches and a social behavior approach. However, the concepts can be analyzed on a second point of view: the first concepts are more physical-specific centric, which means the physical component is relevant and in order to study the event a very specific look is required. A diverse physical agent implies a diverse analysis.

In turn, the final concepts are more social-generic centric, which lead to more generalized perception of disasters, an attempt to find common elements between disasters caused by different physical agents.

In a science committee which discussed the similarities between different types of disasters, Quarantelli pointed out that

"The comparisons attempted clearly showed a conscious belief that trying to perceive phenomena which are not usually grouped together within the same framework, might prevent us from being partially blind in the way it was stated at the beginning of this paper" (Quarantelli, 1981).

In other words, when the researcher sees disasters in a generalized perspective it is possible to notice certain elements that could not be seen if the focus was just in a specific kind of disaster. Quarantelli's statement key word is **framework**. If a framework is designed for disasters in general, that means it could be applied to any sort of disaster with minimal effort.

Quarantelli endorsed a social-generic centric view for disasters, especially when "the problems are divided by time stage, by functions or levels of response"(Quarantelli, 1981). He mentioned Ralph Turner (from the Emergent Norm Theory) who stated "that much of what we know about how people respond to threats and warnings for other dangerous possibilities, is equally applicable to prediction scenarios for earthquakes". On the other hand, that does not imply that the specific study of earthquakes is unnecessary; seismologists still need to analyze earthquakes as much detailed as possible, treating earthquakes as disaster agents. For social and behavioral scientists, though, the best approach is accepting earthquake as members of a more generic class.

The answer for the question proposed at the beginning of this chapter is yes, it is possible to apply the model presented in Section 3 for other types of disasters. However some minor changes must be done in order to use the model properly for an earthquake disaster:

- The threat in a fire incident has physical properties that can be modeled in a simulation as if it was a physical object. Therefore, the fire can be seen, smelled and even heard which implies that the agents can get these physical properties right from the environment and make assumptions on them. An earthquake disaster cannot be turned to a physical object: the whole environment can be felt by the agents. Also, the agent does not measure how dangerous the situation is by looking at the basic physical properties in the same way for a fire incident and an earthquake;

- Although the earthquake is no longer "visible" as an object of its own, it is still visible and noticeable by objects falling and structures crumbling. Also, people still can talk about and discuss their feelings and impressions about the event they are going through, keeping the threat into the communication domain;

- Some basic attributes used by the agents for decision making, such as distance from the threat, are no longer relevant. New attributes and variables must be created, such as the tremor perception. On the other hand, some variables, such as the agent's experience in panic situations, become stronger and even more relevant for the decision making process. Dangerousness and nervosism keep their relevance and usefulness for this simulation;

- The definition of exit as a safe haven remains valid up to a certain level: some buildings have regions that may be used as a safe haven, such as a pillar or under a table. For simplification purposes, the best choice for safety could be remained as the exit of the building;

- Finally, a fire incident could last from minutes up to hours. The simulation presented previously showed a fire incident that last 5 to 6 minutes. An earthquake incident usually lasts only a few minutes not taking the aftershocks into consideration (Bolt, 1973).

The changes mentioned earlier do not imply a physical approach to earthquake disasters because all the collective behavior and panic in crowds' elements (such as the collective behavior stages, the collective mind and so on) remain the same. Besides, these changes can be described as parameters of the simulation and hence the model described in this chapter could be accepted as a framework for panic events.

5. Conclusions

The panic in crowds' phenomenon has been studied for decades by many researchers. Such study is important for predicting and evaluating human behavior patterns in disasters. Although natural disasters are becoming more predictable, their outcomes cannot be easily foreseen. Panic in crowds works as a complex system, which implies that analyzing each individual and element alone does not provide the big picture required to understand the event as a whole. A broader view can notice the behavioral patterns that emerge from the interactions among individuals and it is more suitable for studying hazardous events, such as floods and earthquakes.

Simulating a disaster in real-life is dangerous and unethical. The usage of computer simulations allows the disaster event to happen in a controlled environment with no human loss of any kind. If modeled right, behavioral patterns can be extracted from the panic situation described by the simulation model. Such patterns might help disaster control groups to train people which it will minimize human and material losses. Also, it helps architects, technicians and engineers in designing buildings, rooms and other tools so they have a lower impact on the evacuation procedures during a crisis. Finally, simulations can be used to check and validate new ideas and to propose and check "what-if" scenarios that could be unfeasible to replicate in real-life.

Since panic in crowds is a complex system, a multi-agent based simulation is the best choice to model this kind of phenomenon. This chapter did a historical overview of the collective behavior's studies, since their early ages when collective behavior had a sense of wrongdoing and error up to common, still not institutionalized, social behaviors and the panic in crowds' theories. Everything was bound so further studies could be accomplished and a deeper discussion about the social elements of panic situations could happen.

After that, a simulation model based on the symbolic interactionism and the emergent norm approaches was presented. The model strictly followed the collective behavior formation steps analyzed by the aforementioned approaches and expanded it with computational tools such as expert systems and fuzzy logic. The conceptual model was tailored for fire incidents and a computation model was built, showing that the model can be applied and the fire incident simulation is possible.

Then, a key question was addressed: if it would be possible to use the same model for disasters such as earthquakes. The definition of disaster itself was put into question. As it said earlier, by looking the panic situations as complex systems, a broader view achieves better results than a physical agent focused analysis. Henceforth, the model presented by this chapter could be used for any kind of panic situation, including earthquakes, with minimal adjustments required.

Thanks to the theory and the simulation presented here, new lines of research could be derived. For instance, it would be possible to analyze composite panic situations, such as fire caused by an earthquake, as well as to identify the hazardous and complexity levels of such phenomena which are great pieces of information for authorities and damage control groups so they might create better procedures and allocate resources in critical situations.

6. References

Bolt, B. (1973). Duration of strong ground motion, *Proceedings, 5 th World Conference on Earthquake Engineering*, pp. 1304–1313.

Borgatta, E. F. & Montgomery, R. J. (2000). *Encyclopedia of Sociology*, MacMillian Reference USA.

Brown, L. (2009). Fuzzyj toolkit from the java(tm) platform and fuzzyjess - projects - nrc-cnrc. URL: *http://www.nrc-cnrc.gc.ca/eng/projects/iit/fuzzyj-toolkit.html*

Cohen, L. E. & Felson, M. (1979). Social change and crime rate trends: A routine activity approach., *American Sociological Review* 44(4): 588–608. URL: *http://www.eric.ed.gov/ERICWebPortal/detail?accno=EJ210358*

da Silva, V., Marietto, M. & Ribeiro, C. (2008). A multi-agent model for the micro-to-macro linking derived from a computational view of the social systems theory by luhmann, *LNCS* .

David, N., Marietto, M. B., Sichman, J. S. & Coelho, H. (2004). The structure and logic of interdisciplinary research in agent-based social simulation, *Journal of Artificial Societies and Social Simulation* .
    URL: *http://jasss.soc.surrey.ac.uk/7/3/4.html*

Dimitrov, V. D. & Eriksen, H. M. (2006). How to teach oral ecology using complexity approach?, *Proceedings of the 12th ANZSYS conference - Sustaining our social and natural capital* 1(1).

dos Santos França, R. (2010). *Simulação multi-agentes modelando o comportamento coletivo de pânico em multidões*, Master's thesis, Universidade Federal do ABC.

dos Santos França, R., das Graças B. Marietto, M. & Steinberger, M. B. (2009). A multi-agent model for panic behavior in crowds, *Fourteenth Portuguese Conference on Artificial Intelligence* .

Durkheim, E. (1895). *Les règles de la méthode sociologique*, Presses Universitaires de France.

Freud, S. (1955). *Group psychology and the analysis of the ego.*

Friedman-Hill, E. (2009). Jess, the rule engine for the java platform.
    URL: *http://www.jessrules.com/*

Gilbert, N. & Terna, P. (2000). How to build and use agent-based models in social science, *Mind and Society* 1(1): 57–72.
    URL: *http://dx.doi.org/10.1007/BF02512229*

Helbing, D., Farkas, I., Molnar, P. & Vicsek, T. (2002). Simulation of pedestrian crowds in normal and evacuation situations, *Pedestrian and Evacuation Dynamics* pp. 21–58.

Hsqldb Development Group (2009). Hsqldb.
    URL: *http://www.hsqldb.org/*

Le Bon, G. (1896). *The Crowd: A Study of the Popular Mind*, The Macmillan Co.

Luhmann, N. (1996). *Social Systems*, Stanford University Press.

Mackay, C. & Baruch, B. (1932). *Extraordinary popular delusions and the madness of crowds*, Barnes and Noble Publishing.

McPhail, C. (1989). Blumer's theory of collective behavior, *The Sociological Quarterly* .

Merton, R. K. (1968). *Social Theory and Social Structure*, Free Press.

Park, R. E. (1939). *An Outline of the Principles of Sociology*, Barnes and Noble.

Parsons, T. (1937). *The Structure of Social Action.*

Quarantelli, E. (1981). An agent specific or an all disaster spectrum approach to socio-behavioral aspects of earthquakes?

Quarantelli, E. L. (1975). Panic behavior: Some empirical observations, *American Institute of Architects Conference on Human Response to Tall Buildings* .

Rich, E. (1988). *Artificial Intelligence*, McGraw-Hill, New York, NY.

Ruas, T. L., das Graças Bruno Marietto, M., de Moraes Batista, A. F., dos Santos França, R., Heideker, A., Noronha, E. A. & da Silva, F. A. (2011). Modeling artificial life through multi-agent based simulation, *in* E. A. M. Faisal Alkhateeb & I. A. Doush (eds), *Multi-Agent Systems - Modeling, Control, Programming, Simulations and Applications*, Intech.

Russell, S. & Norvig, P. (2004). *Inteligência Artificial*, Editora Campus, São Paulo - SP.

Smelser, N. J. (1963). *Theory of Collective Behavior*, Free Press.

SwarmTeam (2008). Swarm main page.
    URL: *http://www.swarm.org/*
Tarde, G. (1890). *Les lois de l'imitation: Étude sociologique*, Félix Alcan.
Turner, R. H. & Killian, L. M. (1957). *Collective Behavior*, Prentice-Hall.
Vanderstraeten, R. (2002). Parsons, luhmann and the theorem of double contingency, *Journal of Classical Sociology, Vol. 2, No. 1* .
Zeigler, B., Praehofer, H. & Kim, T. (2000). *Theory of modeling and simulation*, Vol. 100, Academic press.

9

Correlation Between Geology, Earthquake and Urban Planning

Sule Tudes
University of Gazi, Faculty of Architecture,
Department of Urban and Regional Planning, Maltepe, Ankara
Turkey

1. Introduction

Urban planning is the organized planning of the physical environment that the mankind lives by providing the safety in line with the social, cultural and economical needs. The primary objective of the planning is to create healthy, reliable and durable living spaces. At this point, especially earthquakes and their effects in the countries that are located on the seismic belts of the world constitute the primary geologic threshold.

Inadequate consideration of the geohazards and the constraining effects of the geological environment or lack of precaution due to improper projection of the analysis and synthesis results to the planning and the planning decisions give rise to the increase in earhtquake damages. Geological studies aimed at reducing the effects of the ground movements due to earthquakes are of prime significance on the reduction of damage that consitutes the basis of earthquake sensitive planning studies.

Geologists, engineers, architects and planners, in creating the earthquake resistant cities, should determine the geologic hazard processes in advance and for the prevention of hazards turning into the risks and for the reduction of the damage, required precautions should take place in an interdiscplinary work. In this context, the main study is to conduct geologic and geotechnic analyses that will orient the suitability for settlement and land use decisions. The success of the damage reduction work after earthquake is proportional to the scientific base and the accuracy of the decisions.

Geological data that enable the earthquake damage reduction and are analyzed in every plan step seperately should be evaluated in coordination with the criteria of planning and design. The risks resulting from urban texture, building quality, settlement layout and macroform should also be integrated with the analyses and synthesis.

2. Geological and geotechnical parameters in urban planning

Geological and geotechnical investigations that include the details compatible to the planning scale before the planning and design of a city have an indispensible significance in the evaluation of suitability for settlement and land use decisions. The main stage of creating sustainable, durable and safe cities is to carry out natural structure analysis and synthesis by comprehensive investigations (geological, hydrological, engineering geology, geotechnic, seismicity, natural resource analysis etc.) and contemporary scientific methods (GIS, Multi Criteria Decision Analysis, Multi Criteria Decision Support Systems etc.).

Geological investigations are the studies that aim to understand the statigraphic relation of rock and soils and tectonism of the settlement area. Besides, these field data provided by the investigations on the rock and soil of the urban area are to orient the upper scale plans (national and regional scale) and provide a base for detailed geotechnical studies.

Geological studies are the qualitative investigations that mostly cover scientific interpretation, definition and classification. These studies comprise of the tendencies and settlement of upper scale geological structure elements as tectonostratigraphic relations of formations, fault, folding, incompatibility and detailed scale geological structures as layer and of the investigation and the mapping of active fault lines. Geology maps, generally, are prepared in between the scales of 1/100000 and 1/25000 and in planning they serve as a base and guide for the regional and environment plans.

Hydrogeological t models in 1/25000 scale, too, are among the important inputs and natural threshold values of natural structure analyses before the planning. These hydrogeological studies orienting the sustainable planning in the synthesis conducted before the settlement can be listed as the location of aquifer constituting the groundwater, ground water level, direction of motion and its seasonal change, the determination of geological structures (anticlinal, synclinal, fault etc) that direct the ground water and its association with urban structures, surface waters in the areas desired to be opened to the settlement or in current settlements, the feeding and the discharge areas of ground water, natural drainage network detection and its mapping. These geological data, at the city and the basin scale, provide basis for the environment plans which are the planning stages of the settlement decisions. In environment plans, policies regarding spatial distribution of the population, decisions regarding the distribution of infrastructure and settlement units and policies for the reduction of earthquake hazards are developed.

Investigations that reveal the engineering properties of geological units in urban settlement areas are within the context of engineering geology studies. These include experimental studies in the field and laboratory medium rather than observational ones. These studies conducted in rock material and rock mass scale cover the determination of discontinuity properties of the rock masses (its location, number, spacings, discontinuity, infill situation, roughness, etc.) mass weathering degree and mass strength by experimental and emprical methods.

Engineering geology maps provide more detailed information about the soil that the city will rest on with quantitative data and it is an important guide to support the true decision mechanism in both habitability and land use. Generally, prepared in 1/5000 scale, these maps are fundamental basis for the master plans of the planing stage (similarly with 1/5000 scale and on which the macroform of the city is developed) where the decisions on usage such as densities, transportation systems, open green area arrangement, infrastructure, dwelling, commerce are made. When the geological unit in the ground in the urban and new settlement areas has the soil nature, index properties of the ground (grain size distribution, porosity, Atterberg limits etc) and engineering properties (cohesion, internal friction angle, natural unit volume network etc) are determined by in situ (investigation excavations and boring investigation) and laboratory experiments and calculation methods.

Besides, in the urban settlement area, unstable regions and areas that has geohazard (landslide (Figure 1), rock fall-overturn, flood (Figure 2), seismicity, liquifaction, settling-consolidation, carstic cavitations (Figure 3) etc) are analized and their effects on urbanization are investigated.

Fig. 1. An example of landslide in Daly City, California from USGS (US Geological Survey) website. Photographer is unknown.

Fig. 2. An example of flooding in Borçka Town, Artvin, TURKEY. It is taken from http://www.t24.com.tr/haberdetay/54382.aspx. Photographer is unknown.

Fig. 3. An example of karstic space in Yucatan, Meksico. It is taken from
http://www.hackturk.net/komplo-teorisi/287458/cukurlarla-ilgili-komplo-teorileri.html .
Photographer is unknown.

In the cases where geological, hydrogeological and engineering geology studies are insufficient, more subscale maps (1/1000) are used. These studies covering geotechnical investigations, albeit not sufficient for urban design, provide valuable data as the plans showing the cadastre of urban equipment for master plan decisions, city blocks and layout, roads, slopes, bridges, squares, traditional textures. The data for the urban design in building scale are provided by more detailed geotechnical survey on the basis of parcel. On the basis of the parcel, geological-geotechnical investigations, depending on the geologic threshold and the extent of the hazard, can be worked on 1/2000, 1/1000 or 1/500 scale maps as well as on 1/250 scale depending on the extent of georisk within the context of building plot.

3. Earthquake as a planning threshold

The turning point of the transition of the mankind from the rural life to the urban life is the industrial revolution in 19th century. Migration started after this revolution from the villages to the cities has brought several settlement problems along with it. The studies aimed to resolve these problems resulted in the emergence of the urban planning methods and their development. In paralel with the increasing city population, the need for new land for settlement started to increase. This demand of land increased the urban risks by urging to use of the lands unsuitable for settlement and in physical planning the site selection necessitated multiparametered tough decision process. Thus, this caused the development of scientific methods for spatial based analyses statistically and mathematically.

Geological data within the planning discipline, before the planning, are evaluated in investigation, analysis and synthesis stages. These data with suitability to settlement analysis determine the development potential of the city by revealing the geologic threshold and restrictions.

Before the planning all geoenvironmental limiters, geohazards, geological-geotechnical data are evaluated as geologic thresholds. These natural geoenvironmental restrictions, besides the areas of natural hazard (earthquake, landslide, flood etc.), can be classified as cultivated areas and forest lands, water resources, reservations, geological sites etc. These natural thresholds are assessed with manmade thresholds as historical and archeological sites, mania plans, military zones and the habitability analysis are made including urban parameters.

Urban settlements have the tendency to develop with a varying pace depending on the policy of the urban development, economy, geographical features and geologic hazards. On the growth of the urban areas, there exist several natural structure thresholds as topography, soil condition, accessibility. Coping with these thresholds necessitates the analyses and syntheses that are developed by contemporary scientific methods. The selection of the methods of threshold analysis or habitability analysis is based on the number of criteria in the analysis, their quality, self values of the city and made by planners and project group (geological engineer, civil engineer, architect etc). In literature (Dai F. C.,et al.,. (2001), Darvishsefan A. A. et al.. (2004), Jabr, W.M. and El-Awar, F.A. (2004), Kolat Ç., et al.. 2006, Marinoni O. (2004), Marinoni O. (2005), Saaty T. L. (2008)) there developed several mathematical and statistical methods analyzing based on the space to be used in settlement analysis and land use decisions. These methods (Threshold analysis technique, Spatial analysis via GIS, MCDA, MCDSS etc) beyond the sole natural structure analysis, provide the possibility of testing the habitability by correlating physical planning with economical, sociological and technological factors and by considering the current macroform of the city.

Although the thresholds caused by restricting geological environment, depending on the extent of geohazard, sometimes can be overcome by technical precautions, the cost of technique can deadlock the habitability economically. Therefore, threshold analyses should realize the cost analyses of the alternatives besides providing the avoidance of urban risks and be a guide in creating the sustainable and durable cities with minumum cost.

Therefore, the thresholds playing a role in the planning and development of a urban area can be divided into two groups as geoenvironmental thresholds and structural thresholds caused by the macroform of the city and structural features.The first group can be listed as topography, geological structure, hydrogeology, geosites and geoparks, ecology, climate, vegetation, geological properties of urban soil and excavatability, seismicity etc. On the other hand, the second group is the current land use of the city and its infrastructure system.

These geoenvironmental thresholds affect the development and the settlement of the city in different ways. In some cases, these factors can be both advantage and disadvantage for the urban development. Although the hardness and the durability of the rocks in the settlement ground bring extra cost in excavatability, especially for creating earthquake resistant cities and soundness it is a necessary condition. Loose and swampy soils have features that can harm the structure in terms of load carrying capacity, settlement and consolidation problems. Land use decisions as the selection of multi-storey building, medium storey building, low rise building, open green areas and industrial use areas, when considered with ground properties, one area that is not suitable for one use do not necessarily have the risk for some other use. For instance, a swampy area that is not suitable for the construction of multistorey building can satisfy the requirement for open green area arrangement.

As for topographic threshold, with the increase of slope habitability and the cost increases. Rough topography urges the urban design, construction layout, building type and structuring requirements. While 15% of slope in settlement increases the cost, the slope over 30% results in serious technical infrastructure problems. On the other hand, the slope under 5% creates drainage problems.

In an urban area with earthquake hazard risk, earthquake analysis should have the priority and the directive role. The detail and the qualification of the analysis and synthesis before planning exhibit variability from upper scale studies to subscales ones. In urban settlements and development areas, the distance to the fault, the features of the ground, topographic factors, liquifaction requirements, landslides and floods as secondary threats, the ratio of fullness and emptiness, the selection of open green area should be analyzed. Structural order, structuring requirements should be arranged in a way that the effects of probable earthquake are prevented. In order not to have resonance, the interaction of soil and structure and the vibrational periods should be evaluated well. The selection of technology and material that control the building quality should be determined considering the soil condition and seismicity.

In urban design and settlement, prevailing wind direction and insolation are very important. In settlement pattern there should have air corridors to reach all buildings and buildings should be designed in a way that does not interrupt others's light. As for the site selection for the industry, similarly, prevailing wind direction is very important in the sense that spreading malodor to the city and air pollution. Besides, in rainy regions, the risk of flooding should be taken into account and flood risk analysis should be conducted. At the regions under risk, appropriate precautions (correcting the stream beds, leaving the stream beds for open space arrangements rather than opening those to the settlement) should be taken. Moreover, climate properties also affect the foundation type and depth regarding the settlement.

Ecological values are destroyed with the effects of urban development and the natural balance is degenerated. Therefore, in any kind of habitability analysis natural balance should be taken into consideration and the living habitats should be protected.

Urban development areas should be in relation with the current land use. The current transportation and infrastructure system, social equipment, commerce and important centers of the cities should be associated with new subcenter and settlement units. A settlement pattern disconnected from the current city will have difficulty in supplying the needs and developing.

Geoenvironmental and urban thresholds, after evaluated one by one and their priorities and the weights calculated by statistical methods (MCDA, MCDSS, GisVBA, AHP, Grey relation analyses etc.), are superposed with the maps showing natural and human activity thresholds and in final synthesis map the remaining areas out of the thresholds are defined as the urban development directions. Afterwards, the decisions on urban use areas (residential, commercial or industrial) are given in line with threshold analysis and cost analysis.

4. Urban earthquake risk

In the settlements with earthquake risk, for the determination of urban risks geological data analysis is not sufficient alone. Building stock and quality in the urban area and the authentic nature of urban texture are also important factors in the evaluation of the earthquake effects. Therefore, while the urban risk analyses are conducted, all the parameters based on the current settlement quality and features, concentration, equipments, infrastructure and transportation networks should be included in the analyses.

Hazard mitigation studies before an earthquake is the most significant stage of disaster preparation process. In this process, the determination of the primary risks and the corresponding precautions for these risks decrease the life and monetory losses during an earthquake. The first step of the determination of the risks at urban areas is to understand the soil behavior that the city rests on by investigations. Besides, the identification of the building quality of the building stock and the revealing of the soil-structure interaction define the type and the approach of the precautions. New settlements are to be realized under the light of the geological data of the city. The inputs of geological data into the planning and design scale play an effective role in the reduction of urban risks. However, these data should be simple enough for planners and designers so that they are understood and implemented. The accurate use and the synthesis of those data banks are of prime importance in understanding the behavior of earthquakes on urban elements. These data providing inputs for architecture, planning and design shape the city. It is an indispensible necessity that in the creation process of earthquake resistant sustainable cities, geology, planning, architecture and design disciplines work together in a way developing a common terminology.

The risk level of the city changes with the population density, building quality, local ground conditions and distance to fault line. The city is subjected to one single earthquake magnitude and threat, however, settlement units that constitutes the city are faced to different levels of urban risks. The resistance of the settlements that have high quality and earthquake resistant buildings resting on hard soil to the same magnitude earthquake, certainly, will be higher than that of ordinarily constructed areas on problematic and loose soil conditions (Figure 4) due to their geotechnical properties. Therefore, the former will have less urban risks. In other words, in urban areas buildings are constructed with different materials in different structural systems and they can be newly constructed or already completed the economic life. Therefore, at the instant of the earthquake the reaction of the building and the extent of the damage will be controlled by the structural features and the geotechnical characteristics of the ground.

In earthquake prone areas, the effect of earthquake waves on the ground, how this effect is reflected to the building and the reaction of the building to this effect should be clarified in an accurate way by interdisciplinary work. These valuable data obtained by experimental analysis, synthesis and calculations help to the determination of the precautions against urban risks. These precautions can be as strengthening of the buildings or evacuation of weak buildings or abandoning of the settlement area before the earthquake during the stage of hazard mitigation as well as the providing of the transportation of the aids in emergency and constructions after earthquakes through a short and alternative routes and the determination of the regions of emergency action.

Fig. 4. An example of loose soil in Adapazarı after 1999 Marmara earthquake. It is taken from http://avnidincer.8m.com/depfoto.html. Photographer is Eşref Yalçınkaya.

Therefore, urban earthquake risks, essentially, result from the geological, geotechnical characteristics of the ground, tectonism of the region and the relation of settlement area with active faults, soil-structure interaction and topography of the city. These risks, in the soil, can be observed as faulting (Figure 5), settling-consolidation, slipping, liquefaction, land slide, rock fall (Figure 6) etc. The most important risks due to soil-structure interaction are that resonance causing the collapse (prevailing natural period of the building being equal to that of the soil) and soil amplification. The velocity of the earthquakes waves in the soil changes with the hardness and the properties of the soil. For instance, the waves passing through a hard rock mass pass very quickly and the quake is less felt due to the firmness and the voidless nature of the rock while those passing through loose and weak ground pass very slowly filling the voids in the ground and result in the severe feeling of the quake. This behavior of the soil is defined as soil amplification. Therefore, soil amplification factor of alluvial material, which is higher than that of the rock, causes the strong quaking on the alluvial settlement areas with high rate of damage while that of granite will empower the settlement above it.

Fig. 5. An example of faulting on North Anotolian Fault Zone in Turkey. It is taken from USGS (US Geological Survey) website. Photographer is unknown.

Fig. 6. An example of rockfall. It is taken from USGS (US Geological Survey) website. Photographer is unknown.

The closeness to the fault in settlements is not a sole requirement, although it is very important, in the development of urban risks. Certainly, the constructions on the active fault line will feel the quaking more than others. However, the earthquake experiences in the world and in Turkey showed that strong building resting on hard soil could stand

Fig. 7a. An example of landslide. It was devoloped in valley plains. It is taken from http://www.harikasozler.net/img3851.htm. Photographer is unknown.

Fig. 7b. An example of landslide in Laguna Beach, Bluebird Canyon, California .
It was devoloped in steep and high slope. It is taken from USGS website (US Geological
Survey). Photographer is Jim Budak.

regardless of its being on the fault line while the buildings on filled soil far from the fault
line collapse. For that reason, in the spatial plans made in earthquake regions with hard
topography, it should be avoided to settle on the valley plains (Figure 7a) and high slope
(Figure 7b) areas that are prone to landslides.

The liquefaction is a geologic hazard that occurs in the grounds that are cohesionless and
have underground water and if it occurred in the settlement area, it is an urban risk (Figure
8a, b). On the soil that the building rest on, soil-water mixture moving with liquefaction
creates deep enormous voids under the buildings (Figure 8a). That results in the subsidence
or the overturning of the building. Especially, the buildings constructed very closed to the
sea on the sand soil when they are under seismic excitation have the serious risk of urban
collapse and subsidence risk (Figure 8b).

As seen, in the site selection, land use, urban planning and design, geological data are the
main actors of decision process.

During an earthquake, besides the geologic hazards, planning and design errors as wrong
site selection, wrong land use decisions, urban uses off the objective, error regarding the
design, insufficiency of infrastructure, low building quality give rise to the serious urban
risks. For that reason while taking the precautions for the mitigation of the hazard before
earthquake not only risks from the geologic thresholds but also analysis, synthesis and
evaluations regarding all spatial criteria as macroform of the city, design, urban
equipment, concentrations etc should be done and urban risks should be reflected on plan
decisions.

Fig. 8a. An example of liquefaction in TURKEY after 1999 Marmara Earthquake. It is taken from http://www.el-aziz.net/img4381.htm. Photographer is unknown.

Fig. 8b. An example of liquefaction in TURKEY after 1999 Marmara Earthquake. It is taken from http://www.kenthaber.com/marmara/kocaeli/Haber/Genel/Normal/depremde-yikilan-konuta-imza-atti/3d13f1c8-4158-4ce1-b380-13e53de1be21. Photographer is unknown.

5. Earthquake sensitive planning

Tam (2010) defined the earthquake sensitive planning as an integrated planning which aims to mitigate the earthquake risk factor by considering the physical properties and socioeconomic structure of the settlements and which starts from upper scales and develops socioeconomical development policies and supra-national, national and regional plans to further continue to local planning and subscales in which the progressive synergy is assured. (Reference: Deniz Tam)

Earthquake sensitive planning is a planning action that primarily analyzes the earthquake hazard and risks in the planning, prevents these risks and hazards to turn into disasters, internalizes the planning to mitigate earthquake hazards and urban design approaches. The main approach of earthquake sensitive planning is to include the risk mitigation precautions of all disciplines related to earthquake in the planning process for the realization of urban planning that provides healthy, reliable, livable urban environment development.

Earthquake sensitive planning includes the evaluation of geologic hazards and restrictions as risk factors in planning process and their reflection in planning decisions. Within this context, in planning the use of geological data should be assured and regarding the earthquake sensitive planning for hazard mitigation and prevention policies and approaches should be developed.

The process of building earthquake resistant cities comprises the analysis of geoenvironmental natural hazards that can be occurred during an earthquake or after it, the evaluation of the damage assessment and the revealing the corresponding urban mistakes and the conduction of urban risk analyses. Besides, earthquake sensitive planning approach should be developed to eliminate the risk factors due to land use, site selection, settlement pattern and the structuring.

Earthquake sensitive planning is a dynamic action that zooms out urban planning from the spatial design based traditional planning approach, integrates the risk mitigation precautions in the planning process and incorporates the detailed microzoning maps that go beyond the standard geological investigations.

Earthquake sensitive planning involves an analysis perspective starting from the world scale to national, regional, urban and local scale which covers the small settlement units. This perspective bases on the physical, economical and social development and urban risk analysis under the earthquake scenarios.

In every stage of this planning approach, geoenvironmental hazard and risk factor should be determined by geological-geotechnical investigation and microzoning maps and with this geological data analysis there should made feedbacks in every planning stage.

For the reduction of urban risks and hazard, potential development areas with alternatives developed for physical plans by the directive of the geological data should be selected by using the multi decision analysis techniques as well as with the inclusion of socioeconomic analyses.

In earthquake sensitive planning, the interpretations on the analysis of geoenvironmental thresholds and their implementation on the plan are discussed above in the section "Earthquake as planning threshold". In the case where the macroform of the city, layout and socioeconomic development are taken into account, the required action that should be considered in earthquake sensitive approach can be listed as follows:

1. Engineering structures like highway, railway, viaduct, tunnel and construction layout should not intersect the fault line perpendicularly. In the cases where the development

close to the fault is obligatory, urbanization and settlement should be ensured to be in parallel with the fault line (Figure 9).

2. Multicentered development pattern should be adopted and the urban growth should be limited depending on the risk.

3. Population and densities should be arranged in a way to mitigate the risks after earthquake and a balanced distribution should be supported while preventing the increase of the concentrations in one region.

4. The factors that inhibit the socioeconomic development and growth should be resolved and the weight should be given to the process of creating economically powerful and earthquake resistant city.

5. The continuity of the green areas should be provided and in the macroform of the city safe open and empty areas should take their places in settlement units as gathering areas.

6. The transportation network should be built up and for the roads closed after earthquakes, alternative transportation systems should be developed. Within these transportation systems, the shortest route to the areas where the urban risk level is high should be defined to provide means for immediate aid by the analysis of shorthes path via GIS.

7. Technical infrastructure systems should be made resistant to the expected earthquake magnitude by strengthening. Especially, the systems having a vital importance as natural gas pipelines, energy and water lines should be improved against earthquake effects and protective measures should be taken.

Fig. 9. An example of rail way which build on fault zone and is cutted by this zone in Turkey after 1999 Marmara Earthquake. It is taken from http://www.resimkarikatur.com/resim1684.html. Photographer is unknown.

In earthquake sensitive planning, building layout should not be attached. In the cases where it is needed to be attached, story heights should be equal to each other. Different story heights mean different vibration periods. Thus, it may lead to impacts and collapses (Figure 10).

Especially in the developing countries, commerce, urban uses such as residence and social equipments can change their functions in line with the newly emerged needs. This results in the change in the projects of interior design and structural disorder and as a result the building becomes under the risk due to the change in its bearing capacity.

The areas with high geohazard in urban settlements in earthquake prone regions should be left for open green area use.Urban functions should be green buffer zones. These green areas relieve the dense traffic in panic state and easen the intervention as well as being gathering areas after earthquakes.

Fig. 10. An example of collapsed structure because of building design and different construction height after 1999 Marmara Earthquake. It is taken from http://www.haberingundemi.com/haber/Depremin-Simgesi-Bina-Yikildi/80399. Photographer is unknown.

6. Urban settlement site selection and microzoning

Microzoning is defined in a variety of ways by different researchers in the literature (Hays (1980), Sharma ve Kovacs (1980), Nigg (1982), Özçep and et.al, Sherif (1982), Finn (1991)). However, the common point of view of all researchers is that microzoning is to be analyzed in the preparation stage before earthquake to realize the reduction after earthquake while the habitability is to be analyzed especially in the high risk regions by dividing into the smallest subregions. Microzoning maps, depending upon the local geological, seismological and geotechnical conditions, is the mapping of the geohazards of the areas where potential of liquefaction, landslide sensitivity, flood risk, soil amplification etc or combinations of those hazards are seen, as a basis of the planning, the development and the design.

Geological studies and the synthesis of the data used in the planning exhibit a rapid development in terms of directing the planning. Within the framework of this development, geological-geotechnical studies, assessment of the suitability for settlement and microzoning maps provide highly important data to determine the land use and settlement for the planning.

Microzoning maps can be prepared as a base for the 1/100000 scaled regional plans. At the same time it reveals the development direction and the potential of the city by identifying geohazard thresholds for 1/25000 environment layout plans, 1/5000 master plans and 1/1000 tentative plans.

Earthquake resistant building designs advance significantly to decrease the risk of collapse and make the building safe under earthquake loding. However, these designs accompanied by expensive methods and techniques become insufficient in the implementation due to economical reasons especially for the developing countries. Therefore, microzoning maps gains more importance since the selection of a settlement area far from the geohazards will decrease the need for the precautions with high technology.

Microzoning maps direct the plans with an integrated risk approach by evaluating geologic hazards and advantages that are provided by geological-geotechnical investigations with the risks resulted from the constructions.

7. Results

In the planning and the design of new settlement areas and the environment with current settlement, in every stage of the plan geological data with out-of-traditional planning understanding for the reduction of urban earthquake risks should be functionalized in compliance with the objectives.

In the process of creating earthquake resistant safe cities, geological-geotechnical investigations and microzoning maps being an understandable synthesis of geotechnical data play a key role in the integration of hazard mitigation precautions to the planning. However, this geohazard based maps should be developed in complience with the requirements of planning scale and its context. At this point, there seen the necessity of the collaboration of the experts of both geological and planning disciplines.

The planning made in the regions with high earthquake risk should be supported by identifying with the probable earthquake scenarios. In earthquake sensitive planning, the formation of gradual centers system with one main center, the identification of the intensities in correlation with settlement potential, the development of multicentered urban form by preventing urban sprawl are essential.

In the urban areas with high earthquake risk, the improvement of the current plans, the reconfiguration in the required locations and the planning of development areas based on the microzoning maps and probable earthquake scenarios would decrease the probable earthquake damages. In earthquake sensitive planning, the integration of geotechnical parameters of the soil as soil amplification, liquefaction and landslide after evaluation to the planning is of vital importance since these parameters during an earthquake can cause secondary urban risks.

The main factors effective in the distribution of earthquake damage can be summarized as the distance of the settlement to the active fault line, geological structure, local soil conditions, the state of ground water, site selection and land use, population density and distribution, building density, quality, order and design.

As it is seen, the basis of creating a safe and sustainable living space in the urban settlement areas with high seismic risk is the evaluation of urban planning and design, geological synthesis and earthquake analysis in coordination with modern scientific methods and techniques.

8. References

Dai F. C., Lee C. F., Zhang X. H. (2001) GIS based geo-environmental evaluation for urban land-use planning: a case study, Engineering Geology, 61, 257-271.

Darvishsefan A. A., Setoodeh A., Makhdom M. (2004) Environmental consideration in railway route selection with GIS (Case study: Rasht-Anzali railway in İran), Map Asia 2004, Beijing, China.

Jabr, W.M., El-Awar, F.A. (2004) GIS and analytic hierarchy process (AHP) for siting water harvesting. ESRI Proceedings; Available from: <http://gis.esri.com/library/userconf/proc04/docs/pap1539.pdf>.

Kolat Ç., Doyuran V., Ayday C., Süzen M.L., 2006, Preparation of a Geotechnical Mikrozonation Model Using Geographical İnformation Systems Based on Multicriteria Decision Analysis, Engineering Geology 87, 241-255.

Marinoni O. (2004) Implementation of the analytical hierarchy process with VBA in ArcGIS, Computers&Geosciences, 30, 637-646.

Marinoni O. (2005), Adiscussion on the Computational Limitations of Outranking Methods for Land-Use Suitability Assesment, International Journal of Geographical Information Science, 20, 1, 69-87.

Saaty, T.L. (1990) How to make a decision: The Analytic Hierarchy Process. European Journal of Operational

Zhang F., Yang Q., Jia X., Liu J., Wang B. (2006) Land-use optimization by geological hazard assessment in Nanjing City, China, IAEG 2006, paper number 324, Natthingam.

Saaty T. L. (2008) Decision making with the analytic hierarchy process, Int. J. Services Sciences, 1, 1.

Hays, W.W., 1980. Procedures for estimating ground motions, U.S.G.S Professional Paper, 1114, 77 p.

Sharma, S. and Kovacs, W.D., 1980, Microzonation of memphis, tennessee area, A report on research sponsored by The USGS, No: 14.08.0001-17752.

Nıgg, J., 1982. Microzonation and public preparedness: a viable approach, Proceedings of the 3thInternational Earthquake Microzonation Conference, Seattle.

Sherif, M.A., 1982, Introductory Statement of 3 th International Earthquake Microzonation Proceedindgs, June 28-July1, Seattle, USA.

Fınn, W.D.L, 1991. Geotechnical Engineering Aspect of Microzonation, Proc. Fourth Intern.l. Conf. On Seismic Zonation, Vol.1, pp. 199-259.

Tam D., 2004, Çevre Duyarli Planlamanin Ve Deprem Duyarli Planlamanin Bütünleştirilmesinin Sağlayacaği Faydalar, Journal Of The Chamber Of City Planners Union Of Chambers Of Turkish Engineers And Architects, Sayı: 29 Planlama, ISSN 1300-7319.

USGS (US Geological Survey) website.

http://www.t24.com.tr/haberdetay/54382.aspx.

http://www.hackturk.net/komplo-teorisi/287458/cukurlarla-ilgili-komplo-teorileri.html .

http://avnidincer.8m.com/depfoto.html.

http://www.harikasozler.net/img3851.htm.

http://www.el-aziz.net/img4381.htm.

http://www.kenthaber.com/marmara/kocaeli/Haber/Genel/Normal/depremde-yikilan-
   konuta-imza-atti/3d13f1c8-4158-4ce1-b380-13e53de1be21.
http://www.resimkarikatur.com/resim1684.html.
http://www.haberingundemi.com/haber/Depremin-Simgesi-Bina-Yikildi/80399.

Permissions

The contributors of this book come from diverse backgrounds, making this book a truly international effort. This book will bring forth new frontiers with its revolutionizing research information and detailed analysis of the nascent developments around the world.

We would like to thank Sebastiano D'Amico, for lending his expertise to make the book truly unique. He has played a crucial role in the development of this book. Without his invaluable contribution this book wouldn't have been possible. He has made vital efforts to compile up to date information on the varied aspects of this subject to make this book a valuable addition to the collection of many professionals and students.

This book was conceptualized with the vision of imparting up-to-date information and advanced data in this field. To ensure the same, a matchless editorial board was set up. Every individual on the board went through rigorous rounds of assessment to prove their worth. After which they invested a large part of their time researching and compiling the most relevant data for our readers. Conferences and sessions were held from time to time between the editorial board and the contributing authors to present the data in the most comprehensible form. The editorial team has worked tirelessly to provide valuable and valid information to help people across the globe.

Every chapter published in this book has been scrutinized by our experts. Their significance has been extensively debated. The topics covered herein carry significant findings which will fuel the growth of the discipline. They may even be implemented as practical applications or may be referred to as a beginning point for another development. Chapters in this book were first published by InTech; hereby published with permission under the Creative Commons Attribution License or equivalent.

The editorial board has been involved in producing this book since its inception. They have spent rigorous hours researching and exploring the diverse topics which have resulted in the successful publishing of this book. They have passed on their knowledge of decades through this book. To expedite this challenging task, the publisher supported the team at every step. A small team of assistant editors was also appointed to further simplify the editing procedure and attain best results for the readers.

Our editorial team has been hand-picked from every corner of the world. Their multi-ethnicity adds dynamic inputs to the discussions which result in innovative outcomes. These outcomes are then further discussed with the researchers and contributors who give their valuable feedback and opinion regarding the same. The feedback is then collaborated with the researches and they are edited in a comprehensive manner to aid the understanding of the subject.

Apart from the editorial board, the designing team has also invested a significant amount of their time in understanding the subject and creating the most relevant covers. They scrutinized every image to scout for the most suitable representation of the subject and create an appropriate cover for the book.

The publishing team has been involved in this book since its early stages. They were actively engaged in every process, be it collecting the data, connecting with the contributors or procuring relevant information. The team has been an ardent support to the editorial, designing and production team. Their endless efforts to recruit the best for this project, has resulted in the accomplishment of this book. They are a veteran in the field of academics and their pool of knowledge is as vast as their experience in printing. Their expertise and guidance has proved useful at every step. Their uncompromising quality standards have made this book an exceptional effort. Their encouragement from time to time has been an inspiration for everyone.

The publisher and the editorial board hope that this book will prove to be a valuable piece of knowledge for researchers, students, practitioners and scholars across the globe.

List of Contributors

Nuray Balkis
Istanbul University, Marine Science and Management Institute, Istanbul, Turkey

Takeshi Sakaki and Yutaka Matsuo
The University of Tokyo, Japan

Huadong Guo, Liangyun Liu, Xiangtao Fan, Xinwu Li and Lu Zhang
Key Laboratory of Digital Earth Science, Center for Earth Observation and Digital Earth, Chinese Academy of Sciences, Beijing, China

Samvel G. Gevorgyan
Center on Superconductivity & Scientific Instrumentation, Chair of Solid State Physics, Faculty of Physics, Yerevan State University, Armenia
Institute for Physical Research, National Academy of Sciences, Armenia
Precision Sensors/Instrumentation (PSI) Ltd., Armenia

Leszek R. Jaroszewicz and Zbigniew Krajewski
Military University of Technology, Poland

Krzysztof P. Teisseyre
Institute of Geophysics Polish Academy of Sciences, Poland

Guangqi Chen, Yange Li, Yingbin Zhang and Jian Wu
Kyushu University, Japan

Yasuichi Kitamura
National Institute of Information and Communications Technology, Japan

Youngseok Lee
Chungnam National Univeristy, Korea

Ryo Sakiyama and Koji Okamura
Kyushu University, Japan

Robson dos Santos França, Maria das Graças B. Marietto and Margarethe Born Steinberger
Universidade Federal do ABC, Brazil

Nizam Omar
Universidade Presbiteriana Mackenzie, Brazil

Sule Tudes
University of Gazi, Faculty of Architecture, Department of Urban and Regional Planning, Maltepe, Ankara, Turkey